A SNOWBALL IN HELL

A deranged but inventive killer with a wicked sense of humour, Simon Darcourt is busy creating his own celebrity talent show, one that is generating more publicity than its contestants have ever had in their lives. The catch is that those lives won't be very long.

With the police losing the ratings war, they turn to Angelique de Xavia, a cop who has crossed this psycho's path before. However, the police are not the only people who want Darcourt, and ruthless measures are engaged against Angelique as leverage in determining his final destination. Now she's faced not only with tracking down her quarry and spiriting him from under the noses of her fellow cops, but the even more daunting task of ensuring she doesn't end up dead once she's served her purpose.

Scared and alone, Angelique knows she's got a snowball's chance of pulling this off, which is why she's going to need a little magic...

A SNOWBALL IN HELL

Christopher Brookmyre

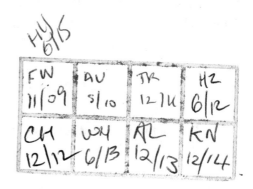

WINDSOR
PARAGON

First published 2008
by Little, Brown
This Large Print edition published 2008
by BBC Audiobooks Ltd
by arrangement with
Little, Brown Book Group

Hardcover ISBN: 978 1 408 41371 5
Softcover   ISBN: 978 1 408 41372 2

Part of the chapter entitled *The Transformed Man*
first appeared, in slightly different form, as a short
story called *Mellow Doubt*, in the charity anthology
Magic, published by Bloomsbury in 2002

British Library Cataloguing in Publication Data available

Printed and bound in Great Britain by
CPI Antony Rowe, Chippenham, Wiltshire

For Marisa

I
The Great Grease-Tailed Shaven Pig Hunt

DEATH NOTICES

Ladies and gentlemen, roll up! Roll on up! Step inside!

You'll find it all in here, you've never seen anything like it, I promise you.

Oh, what a show awaits you, roll up!

You want sick jokes? You want vicarious excitement? You want prurient voyeurism? You want emotion-by-proxy? You want the morally insulated buzz of seeing *other people* behave appallingly? You want sex? You want clashing egos? You want bitching, scheming, clawing, back-biting? You want deceit? You want betrayal? You want violence? You want horror? You want balletically choreographed and spectacularly executed brutality? You want anguish, suffering, humiliation? You want blood? You want death? You want murder?

And you want all of that delivered neatly in a package that lets you lap it up but still feel good about yourself?

Course you fucking do! You're British!

So step right this way! Roll up! Log on! Download the podcast! Tune in! Sky-Plus it, so you can replay the best bits!

It's all here, I tell you. A freak show like nothing you've seen before.

But don't worry, it's perfectly safe. The weirdos, the psychos, the nutters and perverts are all safely insulated on one side of the glass, one side of the CRT, the TFT, the LCD.

Yeah.

3

Your side.

Napoleon really nailed the British psyche with his 'nation of shopkeepers' remark. He didn't merely mean to disparage our modest ambitions and cowering insularity: he truly understood that what went on in those shops defined us more than what went on in our parliaments, palaces or places of worship. His perceptiveness and indeed outright prescience is vindicated in that the quintessential shop he envisaged hadn't even come along yet: the local newsagent, wherein we purchase our beloved tabloids, and over whose counter, accompanied by smiles and please-and-thank-yous and self-satisfied civility, passes the judgmental gossip, envy-driven spite, petty-minded prejudice and that secret delight, that most deliciously savoured hypocrisy, a wee bit of postured outrage.

A nation of shopkeepers, yes, serving a nation of curtain-twitchers: hermetically sealed behind the glass as they spectate upon an absurdly hallowed elite whose lives mean more to them than their own timorous limbos. Never really doing, never really being, always merely looking on, watching other people fight, watching other people fuck. Vicariously living their lives through the attention-gluttonous conduct of the crass and vulgar, and worse, of cyphers just as dull as themselves, but upon whom this latter-day sanctified status of 'celebrity' has been conferred merely by the act of being spectated upon, after which every aspect of their future lives is considered valid and eligible for presentation to the watchers behind the glass.

And listen, listen to that sound this nation of curtain-twitchers makes as it gazes, rapt. It's like the humming of tens of millions of little cicadas in

4

concert, so get yourself close to just one window and concentrate: isolate the sound. Hear it? Yes, there it is: tut-tut. Tut-tut. For disapproval is the keystone: the pitifully unconvincing façade behind which they hide their pallid cowardice, the means by which they try to fool themselves that this emotion they are feeling is something other than jealousy. Tut-tut. It's the talisman that protects them from confronting the truth: that they also have all of the appetites, the lusts and hungers they profess to be disgusted by: they just don't have what it takes to feed.

That's why I've never exactly been inclined to hang my head in shame any time the newspapers called me a monster. I *was* a monster. I am a monster. But let's not pretend for a second that they anything other than fucking loved me for it. I'd have more respect for the cunts if, the next time a serial killer embarked upon his squalid pursuits, one of the tabloids officially sponsored him. They could be honest for a change, have a champagne celebration every time he killed again, in anticipation of their sales going up. Your Soaraway *Sun*: Proud Sponsor of the Summer 2007 Derbyshire Prostitute Slaughter Spree. In tomorrow's *Mirror*: the only *official* coverage of the New Gay Ripper. They could run competitions, like the old spot-the-ball grids you used to get: 'Put your cross in the square on the map showing where YOU think the next mutilated corpse will be abandoned, and you could win a white Escort van, *the* vehicle of choice of several top serial kiddy-murderers!'

Those ridiculously excitable little midgets pulled the head off it every time I pulled off a job. For

5

an industry that practically runs on moral opprobrium, I wasn't merely a tanker of fuel, I was an oil strike, a gusher of the black gold, a gift that kept on giving. They competed to say who hated me the most. I particularly relished the keyboard vigilante types, the ones who called *me* cowardly and wanked on about how much they'd like to be left alone in a room with me. (Careful what you wish for, children.) But deep down, I knew, they were grateful. Christ, look at where they're reduced to getting their moral impetus when I'm not around to provide it. Witch-hunting Jade Goody, I ask you. Almost as much invective spunked out over her as was ever expended on me, not to mention three times as much column acreage, when all she did was be herself—her charming, charitable, literate, intelligent and highly photogenic self—and in the process give the nation a collective showing up. I killed several hundred people, but I think I'd have won a popularity poll against her after Shettygate.

They called me a monster, but they lapped up my every performance. No show without Punch, after all, and my goodness, doesn't this nation of curtain-twitchers love a show.

So roll up, roll up, roll up! Ladies and gentlemen, step this way, and the best part is it won't cost you a thing. The only price is what you're admitting about yourself, and that's no price at all, because we both already knew that about you anyway.

Inside is the reality show you *really* want to see, the star-studded entertainment you're truly craving when you're forced to settle for all that insipid fly-on-the-wall tedium.

6

It's called *I'm a Celebrity and I'm Never Getting Out of Here*.

*   *   *

I really believed I had given up all this sort of thing, you know. A retirement self-imposed largely, I admit, for reasons of self-preservation.

I had a very disastrous and very public failure back on September the sixth, 2001, since which I have endeavoured—most of the time, at least—to maintain the extremely low profile that a widespread belief in one's being deceased affords. The mercy for me, I suppose, was that five days after my snafu at Dubh Ardrain, it was wiped off the news pages and consequently all but erased from public consciousness by events that told me unarguably that the whole game had changed anyway. Talk about burying bad news—New Labour's spinmeisters couldn't possibly compete with the way serendipitous happenstance delivered me my consolation prize. A thwarted terrorist attack on a remote hydroelectric facility in the middle of salmonshire was a big story, especially accompanied by the revelation that it had been the work of the notorious terrorist-for-hire known to police across the globe as the Black Spirit. But there was more: it also emerged that the international contract killer's true identity was that of one Simon Darcourt, Glasgow-born oil-industry executive believed to have perished in the ScanAir Flight 941 bombing over Norway, a terrorist atrocity subsequently attributed to the Black Spirit and thus preceding the Madrid cinema bombing as his acknowledged major-league debut.

Clearly a rather cringe-worthy few days to be me. But once somebody had gone to the bother of hijacking four passenger jets, using two of them to bring down two of the most globally recognisable buildings in one of the most populous and absolutely the most famous city on the planet, killing upwards of three thousand people, before belly-flopping a third airliner into the single largest and most heavily defended building in the known universe for an encore, I realised my own recent travails had been relegated to chip-wrapper status.

I realised also that, even if I had pulled off Dubh Ardrain, it would have been merely a high note to bow out on. September the eleventh would have brought down the curtain upon my stint on the world stage either way.

I was a professional: contract terrorism, some called it; my services available to any individual or organisation who had the contacts to procure them and the budget to meet the price. I'll admit the bottom line was important, but my most compelling motivation in those days was the challenge of pulling it off. I had professional pride, yet it might be more accurate to say, like the Victorian gentleman-amateur, I played for the love of the game. There was no place for me in that game after September the eleventh. The field now belonged solely to the new breed of Islamist fanatics, and they didn't need hired help when they had a host of disposable brainwashed drones to deploy as non-payment-seeking mayhem-delivery systems.

What also became dishearteningly clear in the aftermath was that from here on in, the Incredible Exploding Arabs would be the only show in town.

The USA, having finally endured the indignity of terrorism curling a very large jobbie in *their* fridge for a change, belatedly decided it was unequivocally a Bad Thing. As opposed to an Occasionally Useful Thing when the CIA were trying to destabilise or prop up any given regime in the Middle East or Central America, or a Romantic Misty-Eyed Thing when Noraid were filling buckets to aid 'the struggle' back in the Oul' Country.

I had worked for a variety of, frankly, interchangeable ethnic and political separatists, usually with more money than they had any sense of what they might plausibly ever achieve. Half of them were just playing at soldiers, kidding themselves they were part of some great destiny. The other half knew deep down that their struggle was futile, but in the red-misted tantrum of their frustration, wanted to get a few kicks in at Mummy before they inevitably got their bottoms smacked and sent early to bed. I don't know which constituency was the more pathetic, but I did know that the smarter ones would realise it was now time to cash in whatever chips their armed struggle had accrued and start playing politics with them.

As a theatre of war, terrorism was no longer going to have roles for such minor players.

I did once carry out a contract for an Islamist cell, when I bombed the US Embassy in Madrid. (It was described as 'a cowardly attack via the back door'—or more accurately the back wall, which conveniently abutted a cinema complex.) Back then, so many such groups were disparate, discrete and frequently conflicted, each in dispute with all the rest over whose strain of fundamentalism was

9

the most pure. This particular faction wanted to strengthen their hand at the Islamist nutter table, and reckoned a high-profile attack would be the very dab. They had the funds but not the infrastructure, which was where I came in.

Al Qaeda is usually described as a network, but with 9/11 it was obvious that they had discovered global branding, corporate synergy and vertical integration. They would *not* be outsourcing any more, would not have dealings with anyone who was not a fellow fundamentalist headcase, and had in any event no need for mercenaries when there were thousands of idiots willing to do the work for free.

My skills were not only redundant, they were arguably anachronistic. Any fucking lunatic can take out a target if he's prepared to sacrifice himself to do so. But never mind the skill, where's the fucking fun in that? The real talent, the real *panache* is in being able to pull off the job then get away clean and clear so that you can trouser the greenback and read the headlines.

I kept reading about the daring and mental fortitude of the 9/11 pilots, but point of fucking order: mental fortitude implies a cogent decision-making process, which is patently not going on if you believe that flying a jetliner into a skyscraper is your fast-track portal to paradise, where you'll get to pump six dozen assorted teenage virgins. They weren't mentally strong, they were deluded beyond the point of insanity. And what is it with these fucking people and teenage virgins anyway? Have they ever actually shagged one? I have, more than once, and none of the encounters would appear on my list of sexual highlights. Why wouldn't they

10

rather their paradise be hoaching with women who genuinely know how to fuck? Unless, of course, it's that they're holding out high hopes for that first-time tightness, due to their dicks being so small that shagging an experienced woman is like flinging a sausage up a close.

I digress, however, putting off the contemplation of my own embarrassment. My hubris: the sin of pride. Professional pride is a guarantor of discipline, of protocol, checks, balances and downright fastidious attention to detail. Personal pride, however, is a decadent luxury that I ought to have known better than to indulge during anything other than my spare time. In striving to pull off my greatest achievement, I took my eyes off where they needed to be and suffered instead my calamitous, career-ending fuck-up.

It all came down to a chance encounter with someone from my student years, someone who, like everyone else, believed me dead. Larry the Little Drummer Boy, I called him, aka Raymond Ash. I've had several years to contemplate just how statistically unlucky was this fleeting glimpse we shared at Glasgow Airport, but it's how you respond to the unexpected, even to the astronomically unlikely, that makes the difference between the professional who gets the job done and the whining loser who bitches about his luck. I know that now and I fucking well knew it then. I didn't panic: in a way, it would have worked out better if I had. Instead, I got cocky.

I knew I couldn't afford to let him live, but I should simply have put a bullet in him before he could tell anyone he'd seen me, then disappeared his corpse. Instead I tried to get cute, used him to mislead the authorities, telling myself that this was an integral part of the plan, when it was really an act of reckless arrogance that I couldn't afford. He saw me at the airport, but I made sure he never saw me during his consequent abduction, which I ordered. I guess I was relying upon the cops thinking he was deluded about having seen me, just shaken up by the other things that had subsequently happened to him and pointlessly linking one to the other without any evidential foundation. I let him escape in order that he might lead the authorities to decoy information, and planned to tidy him up as a loose end later, while the cops were busy picking through the wreckage at Dubh Ardrain. But I underestimated him, and certainly underestimated the risk of putting myself in the arena with someone who knew just a little too much about how I think. Foolishly, arrogantly, I handed him the advantage, and he handed me my arse. Him and that cop, the X Woman.

*   *   *

I lost every one of my crew and was lucky— extremely lucky—to escape with my life. Worst of all, my identity was compromised. The police and very soon the whole world knew my real name, my face, and the fact that I did not, after all, perish in the ScanAir bombing.

But maybe, just maybe, there was a subconscious reason I was so reckless. Maybe deep down, as I have often pondered, I knew my cover was irredeemably blown when Larry's eyes met mine, so briefly but crucially, at Glasgow Airport.

12

Perhaps I knew that it was the beginning of the end of that chapter of my life: that Dubh Ardrain was going to be the Black Spirit's farewell performance, one way or the other. Nobody likes to bow out after a failure, but you have to ask yourself whether that failure is a sign that you're not as sharp as you once were, and thus that the time for bowing out is in fact overdue.

The silver lining was that the world believed me dead once more. It was a chance to begin again, to commence a new chapter; though necessarily a quieter and less dramatic one. I had always planned for this. You don't go into any job without knowing how you plan to get out, and furthermore how you plan to get out if that route is suddenly cut off. Just like I had a back-up passport and route out of the UK for if Dubh Ardrain went sour, I also had my short-notice retirement package in place for if I didn't have time to fade away on my own schedule. For years I had not just an identity, but a house, money, a *life* set aside, in stasis, ready for me to step into if it suddenly came time to disappear.

The cops traced me to two houses in France and recovered around eight million euros from various accounts, but they recovered only what I intended them to if ever I had to invoke my emergency escape plan. The assets they got were the larger part of my holdings, but I had always understood that to be part of the price I might have to pay in order for them to conclude that there was nothing more to be found and I wasn't coming back. However, merely being believed dead wasn't going to prove quite the same talisman it had before. There had been only a handful of people who remembered what Simon Darcourt looked like when I 'died' the first time. After Dubh Ardrain—for a brief few days, at least—I was on every front page and television screen in the world. That was why I needed surgery, as were the police's misapprehension to be corrected, my personal apprehension would surely follow. And make no mistake about this: it wasn't being caught by the cops I was worried about.

I have, in my life, once been in locked a jail cell, and once been a guest aboard a billionaire's luxury yacht. It is the latter I am more concerned with ensuring never happens again.

You don't find a professional assassin via a sponsored link on Google. Nor was I out leafleting in ethno-political hotspots offering my services. There was a conduit, a very, *very* powerful conduit, a veritable ventricle through which an engorging volume of blood-stained commerce flowed back then, and no doubt still does. For a while I thought the way the game had changed might have impacted on him too, but it would have been like mere ripples beneath the hull of his vast, gleaming vessel. A man afloat on a sea of blood, always working to ensure that the flow is never stemmed, never missing an opportunity to siphon off some more, and inventively adaptable to an ever-altering environment.

For instance, he had noted the emotional impact of suicide bombings in conveying the strength of feeling that apparently motivated such acts, and the bastard successfully *marketed* the idea by procuring suicide bombers for terrorist causes insufficiently inspirational as to compel any of their adherents to play the human party-popper.

14

However, at that time, suicide bombing was rare enough to still make people stop and ponder the enormity of such a sacrifice. Since then, the impact had been somewhat diluted by what you might call the Gynaecological Proliferation Effect: every cunt's doing it.

He told me the volunteers were 'those who were closest to their god', by which he meant aged and terminally ill individuals ensuring their families were looked after by committing highly conflagatory euthanasia. He may, I realise in retrospect, have been lying about the whole thing. I knew him as Shaloub 'Shub' N'gurath, but he was known to seldom, if ever, give himself the same name to two contacts. He was a man who understood not only the importance of anonymity, but the further effect of concealing himself behind a miasma of myths, rumours, counter-impressions and outright fear. I met him once and once only, before the first job he subcontracted me for, which was when, for all the myths and stories, I was made to understand one thing as fact.

'Professionals do not get caught,' he told me, sharing champagne on one of the sun-drenched, golden wood decks of his ocean-going palace. 'However, I am experienced enough to know that nobody is perfect. Accidents can happen. How is it your own poet puts it, Mr Darcourt? *The best laid plans of mice and men . . .*'

'*Gang aft agley.*'

'Go often wrong, yes. The professional knows when the situation is retrievable and when it is not. If it is not, he knows when to walk away, and he knows to clean up the mess. If you compromise yourself, as far as I am concerned, you have

15

compromised me. If you fear you are contaminated, it is your responsibility to amputate and cauterise before the infection spreads. You find yourself on the run? You do not run to me. If you can stay hidden, stay hidden, but always remember my people will be looking for you too . . .

'If the authorities reach you first, we will get to you wherever you are held. We will break you out if possible, to find out what you told them. If that's not possible, we can get to you inside. There's a lot of things we can do inside too, but ideally we'd bring you back to the boat.'

Did you spot the ellipsis? That's the occasion of some quality curtain-twitcher hypocrisy right there. Yeah, let's just leave it at three little dots, shall we? Because that's the part where he turned on the telly and showed me the hospitality I could expect if I ever found myself on board again. Go on, pretend you're relieved. You can tell yourself it was something best left to the imagination, meaning your sensitive wee self wouldn't even be capable of beginning to picture it. But truth is, you're disappointed that I was coy. You don't want three little dots. You want the gory details.

I saw, on video, a naked man strapped to a steel table, propped upright against a wall, while two guys in plastic coveralls bored holes in every part of his anatomy, using power drills so heavy-duty even these colossi needed two hands to heft them. Let me I assure you that those details remain the goriest I have ever witnessed, but time is marching on, otherwise I'd spare you—spare you? Ha! *Deny* you—nothing.

Suffice to say, it provided a compelling

16

motivation, after Dubh Ardrain, to keep playing dead, and for the most part I did. I didn't entirely keep myself to myself, as any good serial killer's neighbour might attest, but I was careful—usually—to leave no clues suggesting that Simon Darcourt was the author of those deeds. For the truth is, when you possess certain abilities, it is difficult to sit back and watch when you know they would make a difference. Sometimes, a sense of duty prevails. If there are things that truly need to be done, for the common good, and you have the wherewithal to accomplish them, then you could say there is little choice but to step off the bench. There *were* things that truly needed to be done, in as much as there were people who—let's be brutally fucking honest about it—truly needed killing. And don't even begin to argue with me over whether it was for the common good.

However, for me to step fully out of retirement has taken the inspiration of a very special individual, and though I can't truthfully present the common good as a central motive, it does feature as an auxiliary beneficiary. I can't reveal that person's identity quite yet, but I should at this point give a special mention to another remarkable individual with a figurehead role in this affair. There is one overseer, one mastermind of this game, operating at a far higher level than even Shub, who can seek you out when the time comes, no matter how well you've covered your tracks and regardless of where you've chosen to hide. Pale and skinny cunt with a very wicked sense of humour, on whose behalf I have carried out a lot of work over the years, and from whom I have accepted one final contract.

17

All those jobs, all those hits, all those years, and I never once received a death warrant in the form of a dossier in a manila folder like you see in the movies. Yet that was my cue to commence my current project: an A4-size file, containing every detail I needed, complete with ten-by-eight photographs. Small scale, really: single subject, just one individual who had to die, and it would be hard to find anyone who'd argue that this fucker didn't have it coming.

This is bigger than anything else I've ever attempted, and I'll have to cope with all of it alone; no back-up, no infrastructure, no second chances. Thus the planning has been exhaustive, the preparation meticulous and the inventory all but bankrupting, not that that will be a consideration in the end. For this is the most important job I've ever undertaken, as well as being undoubtedly the last. I know that what I am commencing will blow my cover again, and how that must ultimately end, but I know also that it will be worth it. And besides, this time I'll have one hell of an out.

*   *   *

He wakes up in a hotel room, feeling very woozy. His throat's dry as a camel's fart, breath to match. He's got a bastard of a headache, though he can't remember what he may or may not have drunk last night—if in fact last night it was: he feels like he may have been asleep for days. He can't recall anything about the last time he was conscious, or even when that was.

Something feels wrong, something disorientating about his immediate environment. He places his feet down delicately on the floor, at which point he registers that it's a laminate and not a carpet. He squints, his eyes still blurry and not a little sticky from sleep. That's when he screws up his face, registering that the hotel room is not the one he remembers checking into. He looks at the bed, confirming it's a double, checking the far side for evidence of a second occupant, but it's not been slept in, the covers still neatly tucked under the mattress just south of the undisturbed pillows. So that's not it.

He's thirsty like never before in his life. There's a glass of water on a nightstand built into the headboard, and it's as he reaches for it that he notices a folder and a folded copy of a newspaper, *his* newspaper, as well as a remote for the TV. He takes the glass in his right hand and drains it in a parched chain of pulsing gulps, unfolding the tabloid with his left to reveal the front page. His face screws up again and he inadvertently dribbles some of the last gulp from his mouth as he takes in the headline, the picture, the splash bar trailing another story inside. This paper is from last year. What the fuck? He discards it and reaches impatiently for the remote, wincing as the suddenness of his movement exacts a price from somewhere in his skull.

His thumb tries several buttons before the TV responds, coming to life with a high-frequency static ping that causes his eyes to tighten shut for a moment. The TV is showing *Strictly Come Dancing*. He squints at it, checking for the marquee along the bottom with the time and a channel sig that would contextualise the clip as part of a morning news show. There is nothing, just two dancing figures filling the whole screen. That's when he looks to his wrist, then around the room for where his watch might be. He has no idea what the time is. There's light behind the curtains, though his head isn't ready to have the sun blazing into his eyes, so he's not going to open them yet. He paws at the remote again, changing the channel. It shows the same thing. Maybe he just pressed the button corresponding to the channel that was already on. He locates the Channel Up button, gives it a push, then another, and another. There's nothing on but *Strictly Come Dancing*.

He ceases the frantic switching, now paying closer attention to what has been established as being the only programme available on the television. His mouth opens just a little, sign that he's realised more specifically what he's watching, and that it's not the new series, it's a repeat. It's the last series. The series *he* was on.

He doesn't like this. You'd think, wouldn't you, that someone of his make-up would be happy that he's on every channel, but instead he seems to find it disconcerting. Pity. He'll be on every channel again soon enough, for real, but he won't be watching when that happens.

He reaches for the folder now, his hand tentative in its final approach, like he's expecting a static shock from it. He's afraid of what he might find inside. He places it on the bedsheets and delicately takes hold of the top right corner, opening it like it's some centuries-old tome that might disintegrate. He uncovers a sheaf of A4 papers. They're all copies of his columns: mostly photostats, some printouts of the online versions. Paragraphs, individual sentences and isolated phrases are picked out in yellow highlighter. He looks at the pages like they're in Sanskrit or might be some alien artefact. He doesn't seem very reassured by such a familiar sight, familiar words. Starting to get scared now, which is odd, because the clippings all say he's fearless: Darren 'The Daddy' McDade, Britain's Most Fearless Columnist.

He's the scourge of scroungers, pummeller of paedophiles, a one-man border patrol repelling asylum-seekers, the valiant rearguard resistance waging a guerrilla war against political correctness, the toast of white van man and the last advocate of that oppressed minority: the white middle-class heterosexual male. We're going to hell in a handcart, but it's our own fault for listening to the do-gooders and not being tough enough. Tough and fearless like The Daddy.

Some of the blockbusters that he found picked out in yellow in the clippings file:

Muslims are to the new century what Germans were to the last. It's not about a non-representative minority, it's about the majority's eager appetite for what this supposed minority is selling. The Germans bought into this myth of their destiny, time and again, until it was bloody well knocked out of them by John Bull. The Muslims are doing the same thing, with the same visions of world domination, and the same solution is called for. Like the guy who gets too mouthy down the pub, the earlier you give him a slap, the quicker he learns his place and the less chance of him trying it on later.

The inescapable truth nobody likes to bring up about asylum seekers—and I mean the precious few genuine ones—is that if they caused so much trouble in their home countries that they were forced to leave, why the hell would we want to let them start rabble-rousing afresh in ours? If you saw some drunken thug getting ejected from a pub for being out of order, you'd hardly invite him round to your house and tell him to make himself at home, would you?

It's ridiculous, but I'm not laughing.

They keep telling us 'Islam' means Peace. Well, I wish they'd all bloody well give us some.

I've come over all liberal. I've realised the true, genuine plight of the asylum seekers I've previously been so tough on. They're on the run from a regime they can't live under: the regime of putting in a hard day's work for a hard day's pay. Happily for the oppressed, Britain offers asylum from such archaic, non-PC practices. The Land of the Free Handouts is their hope and salvation, a land where they will no more face the threat of having to break sweat.

It's ridiculous, but I'm not laughing.

'Generalisation' is a word liberals use as a distraction to obscure the bleeding obvious. It's ridiculous, but I'm not laughing.

My heart was truly moved this week by the footage of those manacled prisoners in Guantanamo Bay. And the more of the bastards I see manacled, the more I'm moved. Apparently we're supposed to be concerned about their human rights. Seriously. It's ridiculous, but I'm not laughing. If it's all right with the muesli-munchers, we'll worry about preventing these scum from blowing up any more tube trains before we worry about their human rights. The truth is, loath as the hand-wringers are to admit it, you don't end up in one of those orange jumpsuits without a bloody good reason.

As far as I can remember, there has been no end of fat, sweaty, pish-stained, prematurely middle-aged arseholes seeking the cheap route to notoriety and populist approval by acting the keyboard hardman in a tabloid. Some of them were sad enough to believe their own shite, some thought they were just playing the game, posturing for effect or, even more pathetically, playing cheerleader for their proprietor's agenda. They were a pitiful breed of attention-seeking inadequates, little more than drunks shouting at the rain, deluding themselves that they were as tough as they talked. And like drunks, they were largely ignored. Best to give them their space and let them make tits of themselves, because they'll get moved on soon enough, only for an even more revolting specimen to take their place. That was how it was meant to work, anyway, but that was before Darren McDade. That was before

The Daddy inexplicably turned his very loathsomeness into a marketable commodity that made him a regular fixture on TV chat shows, comedy panel games, political discussion programmes—*Question Time*, for fuck's sake—and even, consecrating his loveable rogue status, as one of the celebrity contestants on that fossilised turd from television's Mesozoic stratum, excavated and resurrected to stink anew: *Strictly Come Dancing*.

He had pulled off the audacious cake-and-eat-it strategy of acting like he was a knowing, wink-to-camera self-caricature when he was in TV-personality mode, yet still being able to deliver the hard line straight and true in the next morning's paper. 'String me up, it's the only language I understand': that was his signature hey-it's-all-just-showbiz quip, delivered with what was supposed to convey a good-humoured self-awareness but in practice barely masked a seething contempt for those to whom he clearly felt he owed no apology.

It wasn't a seamless transition between media and between personae, however. Prior to *SCD*, he had generally been confined to later-scheduled shows, so the decision to ratify him as a fitting personality for family entertainment drew quite a bit of flak, particularly coming shortly after his 'Muslims are the new Germans' article had succeeded (and let's not sell him short by suggesting it was anything less than his intention) in getting him reported to the Press Complaints Commission. However, it could be argued that the BBC would be on shaky fucking ground rejecting McDade for a family show while they continued to vomit cash all over that horrible little cunt Danny 'DJ' Jackson. Despite having built a career on pandering to sub-literate bigots, the Beeb were happy to let the Cockney mutant loose all over Saturday teatime, presumably on the understanding that he wouldn't be spewing out remarks about niggers and pakis in front of the kiddies. Thus it followed that McDade could be trusted to keep quiet about poofs and asylum seekers in between bouts of gracelessly hauling his pot-bellied little frame around a ballroom, all the while trying to see down his unfortunate partner's frock.

Besides, it was just spoilsports who were trying to stop him. He was a popular figure, after all. 'He only says what everyone else is thinking,' claimed his editor, Jeremy Seele, 'and in an age when people are gagged by political correctness, he's got the courage to wear his heart on his sleeve.'

He spreads the sheaf unintentionally with his right hand as he leans on it in his effort to climb gradually to his unsteady feet. He's decided he needs to physically explore the room, maybe see if moving about the place shakes loose any sense of familiarity. He grimaces as he reaches a standing position, the ascent of a few feet proving slightly more of a change in altitude than his head can comfortably endure. He stands still for a few moments, like he's making sure he can manage that before trying anything as ambitious as walking. He's dressed in just a pair of pants, one of the leg-holes riding up his crack and exposing his left bum-cheek, a fold of his belly overlapping the elastic stretched around his abdomen. For a guy never done castigating the lazy, he's done well to disguise his athleticism. He turns slowly to his right and looks at the two doors: as with any hotel-room layout, the near one must lead to the bathroom and the far one to the corridor. He can't see a wardrobe or a chest of drawers. Bladder is very full. Time for a pee. Maybe his clothes are in the bog. He sighs and it turns into a yawn, even a bit of stretching going on there. Starting to come round. A pee, maybe a shower. He'll work this out after that. A mate taking the piss, perhaps?

He turns off the TV as he passes, and makes his way slowly towards the first door, gripping the circular handle. It doesn't move. He tries the other direction, still nothing. He gives it a dunt with his shoulder, then has a go at pulling and pushing the handle while twisting his wrist. The door remains closed.

The TV switches itself back on. He turns sharply to look at it, confirming where this sudden sound has come from. He sees himself on the screen, responding to the record low score the judges have handed down.

I saw it, it was a fucking injustice, mate. You were worth at least half a point more than that total.

The locked bathroom is not good, not good at all, and not just because he's realising how much he needs to pee. Could all be a joke, what with the *SCD* on telly, but something's telling him to worry. He goes for the other door. It's locked too. He stands back a second, breathes, tries again, in case he's just got himself in a state. No luck.

His attention turns inevitably to the curtains, the promise of light beyond, an answer to where the hell is he. He walks across the room, round to the other side of the bed, and pulls the right-hand edge of the right-hand curtain slightly away from

26

the wall so that he can just peek behind it. His face contorts in his consternation at what he finds; or rather what he doesn't, that being a window. He hauls the curtains apart and finds only a rectangular white plastic panel, illuminated from behind, and on top of this is a transparent laminated blow-up of another choice quote from his column.

The Daddy. More like the slightly strange unmarried uncle.

The press response plays out as I intended; the better to contextualise the truth when it emerges. Starts off as a bit of a kicker story, the other tabloids having their fun as it comes out that their rival's big-name columnist has gone awol. Lots of 'Jeffrey Bernard is unwell' references when his page has to be filled by some other cirrhotic twat while the paper's senior staff dispatch minions to scour every drinking hole in London in the hope of finding him before the competition do. By day three, Jeffrey Bernard on a bender has become Stephen Fry bailing out of that play. Police have gained entry to his home and reckon his passport is gone. It looks like McDade has left the country. Rumours abound. There is some deal afoot to switch papers. He is in bother with the Revenue. He is under cover, planning to sneak back into the UK by illegal means to show how easily our borders were being penetrated by asylum seekers.

I let it all simmer for four days, then post McDade's heart, pinned to his sleeve, direct to Jeremy Seele. I knew they'd get a DNA match, save me taking the risk of dumping the body somewhere they would find it. Then I wait for the presses to roll.

27

Funny, his paper comes over all coy and unsensational in announcing this unquestionably earth-shattering development. None of the ghoulish slavering over the visceral details that normally colours their reporting of such excitingly blood-drenched discoveries, nor the uninformed speculation about sexual depravities that may have preceded or even led to the grim deed. All the more admirably understated given that I had handed them a head start (or was it heart start?) on the competition, a genuine exclusive to splash. Have I succeeded in ushering in a new era of restraint and sensitivity in the British press's attitude to murder? I bloody well hope not, given what else I've got lined up. Of course, it soon emerges I had no reason to worry. The rest of the press fail to suffer the same sudden bout of squeamishness, filling their boots with undisguised alacrity. Naturally, with no details other than what the police and McDade's employers are prepared to reveal, they have to fill the void with two things: constructing their own competing versions of the late Daddy's life, and speculating blindly about the manner and motive behind his death.

Very disappointingly, it takes two days before any of them show a sufficient knowledge of the history of their own profession as to recall that vital organs had been posted to the press during the hunt for Jack the Ripper. And again, they don't quite warm to the theme as much as one might expect. A murderer putting the press front and centre while invoking the most famous serial killer in British history? They should have been spraying their pants at the prospect of further atrocities. Amazing what a sobering effect it has on them to think that their profession might be the one being preyed on for a change, and that it is a different kind of whore who might find himself giving new meaning to the phrase 'newspaper hack'.

Yes, they might be a bunch of two-faced, back-stabbing twunts, but they can still pull together when facing a common threat. Spirit of the blitz and all that, gorblessem. Hence beaucoup quality po-faced revisionism in all printed recollections of the departed, facilitated by the impressively audacious hypocrisy only the British tabloids can truly pull off.

People will no doubt say I was sadistically cruel to the man, but for a few days at least, I turn him into a fucking hero: from a pot-bellied, beery-chopped little bigot into a paragon of British values and a legend of journalism. Oh, the humanity. What a tragic loss! All that charity work, apparently. (No specific charities are named, but that is surely down to his colleagues respecting his well-known modesty regarding such matters, rather than there not actually being one with any record of him having lifted a finger for.) A wonderful colleague to work with. Admired by his readers, feared by his enemies. And as for those accusations of minority-bashing and that slight blemish of being referred to the PCC, where it isn't Photoshopped out of the picture altogether, it is justified as stout-hearted, fearless defence of his principles. He called it like he saw it in the face of the PC backlash; what you saw was what you got, didn't care if his views were fashionable or popular (yeah, right); unflinchingly endured the slings and arrows blah blah cliched blah. His paper even manages to wheel out his local Asian shopkeeper

29

so that they can play the 'look, some of his best friends were darkies' card.

A man of principle, no matter what else anyone might say. That is the consensus. Love him or loathe him, you had to respect the fact that he stood by his beliefs. The phrase 'he wore his heart on his sleeve' is entertainingly conspicuous by its absence, but his journalistic heroism is rendered so unquestionable that they are all but erecting a statue to the prick.

For a few days.

I hold off until it has reached what I estimate to be its zenith before leaking the video on to the web. I time it for when I know the final editions have gone to print, so that the press are effectively a whole day behind the biggest story in the country—cutting them out of the news cycle and underlining their growing obsolescence.

It's funny watching McDade's newspaper trying to get the file-hosting sites to delete it. The same paper whose online edition runs links to the Saddam execution and hostage beheading videos. The best part is witnessing this endangered species of a medium doing a King Canute act in the face of the technology that is superseding it. There are no exclusives and no injunctions in cyberspace. Once the file is on the web, I know it will proliferate exponentially. Christ, b3ta.com are already Photoshopping image-captures from it for sick laughs before the first print response will hit the streets.

So let's go back to that video now, shall we? Okay, where had I paused it? Oh yes.

McDade stands back so that he can read the text.

How did we ever lose our moral grip to the extent that the muesli-munchers convinced us we need to be 'understanding' of murderers, rapists and other scum? Can you imagine trying to explain to school children that they mustn't step out of line because if they do, they'll get taken to one side and given counselling? Better still, can you imagine that classroom about half an hour after that particular lesson has sunk in? Now fast forward to those same kids all grown up. 'I felt like shooting this jeweller and ripping off his store, but the threat of being sat down with a psychologist stayed my hand.' Don't do the crime if you can't face the sympathetic ear.

It's ridiculous, but I'm not laughing.

Criminals need to fear prisons. Prisons need to punish.

Who would you rather find yourself locked in a room with? A thug who knows anything he does to you will be paid back in years of hard time, hot sweat and cold fear, or one who knows the scariest thing he'll face as a consequence is a weekly session with a *Guardian*-reading sociologist called Quentin.

He stares at the panel for a few seconds. He recognises the words, doesn't have to read them all; he remembers. But he's starting to ask himself why he's seeing them in lights, and more pressingly, what significance they have to what he now understands is his captivity.

I let him ponder that for a while, long enough

for him to try both doors again. His bladder's really insistent by this point. The doors remain locked. He tries to shoulder the one he reckons leads to the bathroom, then when that fails to budge, takes to thumping the one he has assumed to be the exit. There is, of course, no bathroom; and for this piece of excrement there is, most definitely, no exit.

At last, he speaks.

'Hello? Hello? Okay, come on, you've had your joke. Who's there? Come on, for fuck's sake, I need to piss.'

There's an edit on the video at this point, black screen captioned: 'Time passeth', with a digital clock fast-forwarded to show twelve minutes elapsing. It fades up the action again to show McDade pacing the carpet between the two doors.

'Listen,' he calls out. 'I need to piss, and if you don't let me out, I'm pissing on your fucking carpet, all right? I'm warning you.'

Yeah, he's giving out warnings at this point. Still the Daddy.

'Okay, that's it.'

With which he stands in one corner, gets out his piggly-wiggly little tadger and urinates copiously on to the laminated floor.

Another 'Time passeth' caption. This time six hours.

He's rather fraught when we see him again, sitting on the bed, biting his thumbnail and looking like he's thinking seriously about the comfort sucking it might once more afford.

A voice suddenly booms through the room, causing him to shudder and stiffen.

'Darren McDade,' it says. He looks around for

the source, but it seems like it's coming from all directions at once. 'Are these your words?' I ask, and cue the lightbox to flash.

'Who's there?' he asks, getting to his feet.

'Are these your words?'

'Well of course they're my words, you know they're my cunting words. Who the fuck is this?' He's trying to sound defiant; there's anger in there, but he can't quite disguise the fear.

'Read them aloud,' I demand.

'Listen, whoever you are, you've had your fun. But I'm warning you, I'll be having mine. Do you know who I . . . I mean, you should bloody well remember who I am. I've got a lot of friends in the police, cops fucking love me. And if you don't let me go right now, you'll be falling down a lot of stairs when they get you in custody, do you understand?'

Six more hours passeth. The floor getteth wetter.

'Read them aloud,' I request again.

He reads the passage perfunctorily, doesn't exactly sell it. But who needs to be an orator when the words themselves carry such weight. Oh, and when the words espouse such sincerely held principles.

The TV switches itself back on, no longer showing *SCD*. Instead, it displays CCTV footage of a tall, heavy-set but muscular, shaven-headed white male, stripped to the waist the better to show off a body, arms and scalp so oversubscribed with tattoos as to resemble a public lavatory wall bearing at least a decade's graffiti. Hard to gauge his age: could be anything between thirty-five and fifty, but however long he's lived, he's clearly lived

33

hard. Very, very, white-supremacist-prison-gang-crystal-meth-biker-crew hard. He's pacing a short, blank-walled room, restless, twitchy, apparently muttering to himself, though it's a silent feed.

McDade stares at the screen, questions racing through his mind. His captor? A fellow prisoner?

Then the image switches to an identical blank-walled room, this time accommodating another tall, imposing-looking figure: an athletically built black male dressed in a pair of grey jogging trousers and a tight t-shirt. Looks early thirties, maybe more, maybe less. There's still youth in his face but a tiredness in his eyes. He is leaning against a wall, arms folded. He looks calm, patient, maybe just a little calculating.

McDade gets a look at him too, then the screen splits vertically to show both of them simultaneously. There are time codes in the bottom-left corner of each image: date, hour, minute, seconds. This is now.

Time to play.

'Mr McDade,' I tell him, 'I have gone to a lot of trouble—a *lot* of trouble—to arrange this little rendezvous with these gentlemen. The stakes are very high—well, your stake is anyway—so I would advise you to listen most carefully. The two men you can see are both visitors from the USA. Both men are convicted murderers. Both men were jailed for horrifically brutal gang-related killings. Both men served more than fifteen years for their crimes. Of more immediate pertinence, both men believe they are about to enjoy unfettered, unchaperoned and uninterrupted access to a secret, anonymous police informant whose testimony was crucial to their convictions. Both

34

men are also under the correct impression that there is very little chance of anything they do to you ever coming to the attention of the authorities.'

With this, the TV switches to a news report updating the ongoing search for the missing journalist, before flashing up a couple of scanned pages from the early press coverage.

'Nobody knows where you are, Mr McDade.'

McDade says nothing, but casts a glance towards the doors, then back at the screen with an aghast scrutiny.

'Now, here's where it gets fun. As we've seen from the previous footage, you're a jolly good sport when it comes to game shows, so I've lined up a little challenge for you—and don't worry, this one won't involve any dancing. One of these men spent the last five years of his sentence undergoing a voluntary rehabilitation programme. The other did nothing but hard time in one of America's most notoriously brutal penitentiaries.

'To get out of this room, you're going to have to pass through a chamber containing one of these convicted murderers. If you can get past the man who is waiting for you beyond the door of your choosing, you are free to go. To make it an informed choice, I will tell you that the door on your right, which you previously believed to be the bathroom, leads to Mr Rehab; and that the door on the left, which you previously believed to be the way out, leads to Mr Hard Time. I will not tell you which of these two men is which, but I will remind you that both have been informed that you were instrumental in their convictions.'

He stands frozen, his focus alternating between

the doors and the split-screen view on the TV like he's watching a tennis match. He stands like that for a couple of minutes, then sits down on the edge of the bed and folds his arms.

'Nah,' he says. 'This is a wind-up. It's a fuckin' wind-up. It's that cunt Beadle, Candid fucking Camera or something.'

That's when he hears the gas beginning to pump into the room. He bolts upright again in response to the sound, which is coming from a vent high on the same wall as bears the light panel. A quick sniff tells him it's the North Sea's finest.

'The cubic volume of this room and the lack of ventilation means that the concentration of gas in the air will become lethal in approximately six minutes, Mr McDade. That's how long you have to choose which door you believe offers you the safer route out.'

He instinctively puts a hand over his nose and mouth. It's not going to save him, but it proves he understands there is no option to abstain. He looks at the screen again. The biker-nazi guy is prowling, unable to contain himself. If there was a sofa in there with him, he'd be pulling the stuffing out of it. The black guy remains impassive, occasionally shifting which leg most of his weight is resting on, biding his time.

McDade takes a few steps across the floor and stands roughly midway between the doors. His face contorts. He can *taste* the gas. He coughs.

'Which door is which?' he says hurriedly. 'Okay, I'll choose, but I've forgotten. Tell me, please, which door is which.'

I give him a while to stew, long enough for him to look up at the ceiling as though I'm floating,

36

disembodied, above him, and scream: 'WHICH DOOR IS WHICH? TELL ME!'

'Bathroom for rehabilitation. Exit for hard time.'

He takes one last glance at the screen, then the man of sincerely held beliefs and enduring principles lunges for the bathroom.

That's where the first video ends. The following interlude I also recorded for posterity, but for a number of reasons have opted not to share with the general public.

McDade emerges into not a bathroom but a narrow passageway, about ten yards long. He pulls the door closed behind him to cut off the gas, at which point a door at the far end swings open. He stops dead at the sight of this, bracing himself for what might emerge through the aperture. Nothing does. He takes a moment to clear his lungs of the gas and gulp down some uncontaminated air, then begins to walk tentatively forward. He can see that the door doesn't lead to either of the narrow chambers he had been watching on TV, but to somewhere more expansive. He can see one wall, maybe twenty yards distant, but the light is low, the space beyond gloomy and windowless. He pauses at the door, angling his head to try to see what might be to his left or right, then steps forward with a hilariously ginger gait. Once he has cleared the doorway, I emerge from the shadows.

He flinches, throwing himself back against the wall to the left of the doorway, head down, barely daring to look directly at me.

'Look, whoever you are, you've got to believe me, I've never been a police inform—'

'I know,' I tell him.

At this point he looks up enough to recognise that I am neither of the men he saw on screen.

'I lied about what was waiting behind the doors. What you saw was just stock CCTV footage. I have no idea who those men were. And as you have discovered, both doors lead to the same place. But it's a lie redeemed if it causes a greater truth to emerge, wouldn't you agree? Rehabilitation or hard labour: it was a no-brainer really, wasn't it, which one made for a safer route? That's why I assigned to the bathroom the option I knew you'd choose: because you're full of shit. I was telling the truth about the rest, however. To escape, you merely have to get past the man you found on the other side.'

'Who are you?' he asks.

I should add that I'm wearing a mask at this point. This is, among other reasons, so that he believes my wish to protect my identity implies that I ultimately intend to let him live.

'There is another selection from your greatest hits on the floor just in front of you,' I tell him. 'Read it.'

He looks down, notices the sheet of paper lying on the concrete. He scuttles forward to pick it up, keeping his eye on me as he does so, as though afraid I'm going to rush him.

The paper bears a blow-up photocopy of one sentence:

The greatest tragedy about the death of
Simon Darcourt is that none of us will ever
get to spend ten minutes locked in a room
with him, without his guns, his C4 and his
henchmen.

He looks up, his jaw hanging, his breathing suspended.

'Your ten minutes start now.'

Video Two starts by tracking out from McDade manacled to a chair, the chair itself bolted to the floor. He's got a sack over his head and is wearing an orange jumpsuit. I take the bag off and step out of shot. His face is bloodied and bruised, though it hasn't had time to really swell up yet. He looks ahead and immediately starts fighting against the restraints, throwing himself back in the chair, but it isn't going to budge. Out of shot is a steel table, upon which he can see a car battery connected to two crocodile clips; a basin of water floating a large sponge; a rope; an oxyacetylene blowtorch beside a welder's mask; an unfolded canvas wrap accommodating a selection of surgical instruments; a pair of bolt-cutters; and finally a large hypodermic needle attached to a drip-line next to two bags of saline solution.

He pisses himself. It's ridiculous, but he's not laughing.

'Hope you like the togs,' I tell him. 'I'm sure you're getting a bit fed up hearing your own words, but as you put it yourself, you don't end up in one of those orange jumpsuits without a bloody good reason.

'Now, listen carefully, because this is about securing your future. If I was to let you go, you would no longer be able to live safely in this country. I'm prepared to let you leave, but if you return, I will kill you. Do not even begin to kid yourself that the police will be able to protect you.

I've abducted you once and I can do it again. If you remain in the UK, I promise I will kill you, but not before an ordeal beyond the worst horrors of your imagination. Do you understand?'

'But . . . but . . . I'm just a journalist. Just a pundit, a talking head. I'll change what I'm writing, just tell me what you want.'

'I've told you: I want you to leave. I don't want you in the country any more. And I'm not merely taking away your right to express your . . . well, in light of your earlier decision we can hardly call them beliefs any more, can we? Let me rephrase: I am not merely taking away your right to publish your moronic populist drivel, Mr McDade. I am taking away your right to live unpersecuted, to live unoppressed by constant fear. If you remain in this country, I will come for you once more. You'll never know when, you'll never know where. You'll just know that if you stay, you will be apprehended in darkness, once again, taken to a place of confinement, and killed in the slowest and most painful manner I can dream up. Do you understand?'

He nods like a Parkinson's sufferer.

'I understand. I'll leave, I promise. I'll leave immediately.'

'You'll flee to another country?'

'Yes, I swear.'

'Seek asylum somewhere abroad?'

'Yes.'

He states this effusively, still nodding, but you can see it in his eyes the moment he realises what he just said.

'So, to recap, Mr McDade, you're now an asylum seeker who believes in rehabilitation for convicted criminals.'

He raises his head, still quivering but perhaps daring to believe he's heard the punchline and the joke is finally over. He's almost right.

'Sorry, I lied again, about letting you go. Another lie redeemed by the truth it revealed. I'm going to kill you now.'

'Jesus, please,' he blurts, imploding into whimpering sobs.

I take a walk towards the table, causing him to throw all his strength into rocking back and straining at his bonds.

'Don't panic,' I reassure him. 'I'm not going to torture you. That would be inappropriate. Instead you will be the author of your own death; or at least the author of the method. Your words once more: "String me up—it's the only language I understand."'

I lift the rope from the table and carefully begin tying it into a noose. The knot accomplished, I turn back to McDade.

'I believe it's traditional for the condemned man to be offered a last meal. Would you care for some muesli?'

GLITTERBALL SHARDS (I)

Zal's hands are cuffed, tight about both wrists, the jangling links of solid steel looped behind his back around an upright column. He is on his knees, his bare feet similarly bound to the immovable pillar. The steel is warm from the hands that fastened it, moist now with two people's sweat. His arms are stretched behind him, his back tight to the column, his posture cramped and contorted.

His captor has retreated from sight. Zal is now isolated, hidden from any observer, cut off from all intercession. The bounty hunter, though unseen, remains mere yards away, rapt in his vigil.

Zal allows himself a moment to contemplate precisely how his situation must look from that bastard's point of view: one man relishing the other's inescapable captivity, blissfully unaware that he has the picture back to front.

Zal smiles and whispers to himself: 'Alakazammy, stairheid rammy. Suffer, you prick.'

COVER STORY

'Allah hu akhbar, Allah. Allah hu akhbar, Allah. Hay-ye alas-slat.'

He hears the muezzin's voice, calling out like a siren from the minaret. There's times of day when it's swallowed by the traffic on Boulevard Aristide Bruant and you can only hear it if you're already in the building; not the most effective call to prayer as you need to already be in the mosque to hear it. But this morning, this Sunday when the shops are closed and the Christians make their token, largely grudging weekly display of faith, it's easy to imagine it sounding out across the rooftops, heard for miles around, echoing along empty streets, rapping against the closed shutters behind which most natives are practising their true faith, that of mattress-worship.

He's sweating, though the air is cool. He looks again through the window down to the haram. That's where the call to prayer is most audible even at the noisiest of times. There are dozens of people hastening through the courtyard, but instead of hurrying to prayer (*hay-ye alas-slat*), they are evacuating the mosque, scattering anxiously out towards the street.

'Ash-hadu an la ilaha illal-lah.'

The muezzin still calls them to prayer, though this is neither a renewed appeal nor an act of faith that they will return: the voice is on an iPod plugged into a speaker system atop the minaret.

There are some screams from women, panicked cries from men. Voices sound out in both French

43

and Arabic. They are urging each other to run. The word 'police' bounces about the courtyard like a teargas grenade.

Other people have noticed the commotion and begun streaming through from the madrassa, heading for the courtyard and out. He is suddenly swimming against the tide. People urge him to turn and leave, like he is a heedless child.

'Gunfire!'

'Police! They have guns!'

'There is talk of a bomb!'

'The cursed jihadis have brought this down upon us. They will not rest until they have made martyrs of us all.'

He lets them pass, buffeted slightly by the occasional outstretched appealing hand, standing his ground and waiting for the door to the madrassa to clear.

'It is an outrage,' he spits, to none of them in particular. 'They would desecrate our place of prayer! An outrage. A rape.'

There is one person not hording through with the rest of the evacuees: a woman in a niqab, standing helplessly apart from the charge. Hands are offered to her too, but she shrinks back as though in panic, even turning her back to prevent further entreaties. She must have come from the kitchen, perhaps. Maybe she doesn't understand what is being said: could be she speaks only Farsi, or Urdu, her husband preferring her not to learn French. And where is her husband? Perhaps that is why she is so helpless.

There is a louder swell of voices from just outside as the last of the evacuees depart the passageway. He sees men throw themselves back

44

against the wall, a flash of black uniform, the glint of grey. Gun-metal grey.

Then one more body moves aside and his eyes meet those of the policeman.

Recognition is instant.

'Akim Hasan. Arrête.'

The sub-machine gun is raised to shoulder-height. He freezes for quarter of a second, then another evacuee, a student, staggers inadvertently into the policeman's view before ducking down with his arms covering his head.

It's long enough. He launches himself to the left, out of the policeman's line of sight. The door to the madrassa is clear now, but he won't make it, not without putting himself directly into view. However, the woman in the niqab, in her erratic helplessness, has moved to within a few feet of him.

As the policeman steps through the archway and into the passage, he grabs the woman around the shoulders, putting her body in front of his, then from a pocket produces a handgun and places it against the black cloth at the side of her head. Others might call it cowardly; in his mind it is entirely justified. If she dies, she will go to paradise, but on this earth, anything is a valid weapon against the *kuffar*.

The policeman stands fifteen yards away in a trained stance: knees slightly bent, the machine gun held at shoulder-level, his eyes looking along the sight, which is now the second weapon to be pointed straight at the woman's head. He is in full body armour and helmet, a thin boom microphone in front of his lips.

'Drop the gun, Hasan,' the policeman says, in

45

French. His voice is hurried, anxious. 'Let the woman go. We've got police everywhere.'

'You drop your gun. *Then* I'll let her go.'

Hasan. Akim Hasan. That's his name.

It used to be Jean-Marc Danticathe. He's been radicalised. It's clear from his Arabic. It's stilted, too formal, sticking out like a freshly erected wooden post in an old fence: unweathered, stiff, clean, missing the rough edges that characterise long-term use.

The policeman is looking at the woman, occasionally getting to glimpse one of Akim's eyes as he peers past his human shield.

'Put down your gun, please, officer,' she appeals, also speaking in French. So she *could* understand what was said. Why didn't she run? Scared? Lost? Hysterical? Doesn't matter now: he's just grateful that she didn't. 'Please, officer,' she repeats. 'As you said, there are other police here.'

The policeman retains his stance. His eyes narrow briefly: a message in his earpiece, perhaps. Then he swallows and says: 'Okay. I'm putting my gun down. Don't hurt the lady.'

'You put it on the floor and then slide it towards me,' Akim says. He takes a step back, pulling his hostage with him.

The policeman crouches to his knees, turning the gun to present it sideways all the way to the floor, his focus always remaining on what he can see of Akim and the two eyes that are the only part of the woman visible through the slit of her niqab.

He switches on the safety and gives the machine gun a stiff push across the flagstones. It comes to rest at the woman's feet. She turns her head to look out of the window. Akim puts a leg around her and places a foot on the stock of the machine gun, intending to drag it behind him as he backs himself and his hostage closer to the far exit, towards the madrassa.

As he does so, the woman's right hand suddenly grips his wrist while her leg shoots up beneath his outstretched thigh, a microsecond before she pivots on her standing leg and sends her left hand, fingers outstretched, into his throat. She then vanishes from under him, reappearing at his back, her grip on his wrist somehow retained throughout the manoeuvre in a way that has drastic ramifications for the connectivity of this particular upper limb.

His face hits the flags with a crack almost as loud as the one accompanying her breaking his arm. Then she is kneeling on his back, pointing his handgun into the base of his skull, while checking the window again.

He's screaming, curses and hatred spewing out amid the cries of pain. The infidels are getting it pretty tight, and the Jews aren't coming out of it too well either. She wonders if it would hurt more if he knew that the Krav Maga techniques she just used were devised by the Israeli security forces. He doesn't realise it, but she just saved his life. Another foot backwards and one of their marksman would have had a clear shot.

They need him to talk, but that's not the principal reason she wanted to make sure he was taken alive. Everybody deserves a chance to change, to repent and to make amends. Dead people could do none of the above. Hope had to prevail, no matter how unlikely. One day, some Islamist firebrand leader was going to have an epiphany and realise that a book largely concerning seventh-century tribal squabblings was a less-than-sturdy basis for advocating mass slaughter. One day. The planet had a few billion years left before the sun went supernova and consumed the solar system, so there was probably enough time, but it might be tight.

She stands away and lets the policeman recover his weapon. Dumarque, his name is.

'You all right?' he asks.

'A little hot under the canopy here, but yes.'

Dumarque smiles.

*　　　*　　　*

Aye. There was, upon reflection, something to commend the much-maligned jilbab-and-niqab combo, to say nothing of the burqa.

Sure, it was harder to think of a more complete symbol of absolute female subjugation than forcing them to dress like a free-standing Victorian changing booth rather than the human being that might be found inside it. Sure, it showed disgust for the female body, an insulting mistrust of male self-control and a bridling hatred of human sexuality in general. Sure, it was only a kick in the arse off of locking your woman away in a black box like some possession that had no autonomous will of its own, there only to be used when required and then securely stored again. Sure, it denied the woman the right to any pride or pleasure in her form or appearance, the right to express any individuality and the ability to indicate anything through non-verbal means other than whatever she might be able to convey by making her eyes go

48

skelly. Sure, the Islamic fundamentalist version of *Big Brother* would be a bastard to follow unless the women all wore really big name-tags. And sure, the only way a culture could have been more contemptuous in devising a garment so sadistically unsuited to its own prevailing climate would have been if eskimoes sent their women out in thong bikinis. But . . .

When it came to undercover surveillance of Islamist radicals, it was simply *de rigueur*, darling. It was like wearing the One Ring. For a start, nobody knew what you looked like underneath the tent. Nor could anyone guess what else might be concealed beneath its billowing folds. A wire? A pinhole bodycam? Hah! You could fit two boom microphones and a Dolby stereo pan inside and still have room for a high-def digital video camera up each sleeve. But its true Klingon cloaking-device efficacy lay in the fact that merely wearing it marked you as toeing the line, knowing your place and posing no threat. Invisibility rendered through irrelevance. You were, after all, a woman; and not just that, but a woman who was under control and who knew how to behave.

She had been amazed at how unguarded some of her surveillance targets could be, the candid matters they had talked about while she and other similarly (similarly?—that would be identically) attired women were in earshot. It was as though these guys thought they spoke a different language, or were discussing matters the women couldn't possibly understand, but just as likely their simple assumption was that the women knew it was none of their concern and would obediently disbar it from entering their heads.

It was a phenomenon far from unique to radical Muslim culture. Her mum had once sat in the same train compartment while four suits discussed the finer points of the boardroom coup they were planning. It didn't occur to any of them that she might be on her way to the same AGM and annual conference as themselves. But when the radical philosophy you had taken it upon yourself to impose upon the entire world ascribed women a role somewhere between indentured servant and dumb but faithful family pet, the effect could be a little more pronounced. Plus, in the world of the niqab and the burqa, you could pitch up anywhere without fear of someone saying, 'Hey, I haven't seen *your* face round here before.'

In many ways, it had been the easiest surveillance work she'd ever done. The hardest part had been immersing herself in another religious culture, having spent enough years trying to scrub off the emotional detritus of the one she was brought up in. It was depressing to think how much time she must have spent hanging around mosques, considering she was already bitterly mourning every last second of her life that had been wasted inside Catholic churches. All those dreary Sunday mornings throughout childhood, surrounded by equally miserable people too inured to the mindless tedium of it to be asking themselves whether what they were doing made any sense or why they were there at all when the experience gave them nothing in return except the odd skelf in their arses. It was only when you stepped outside of it that you could truly take in the scale of the absurdity. The few times she had returned to a church service in the years since had

50

been for weddings and funerals, and what she had seen from this refreshed perspective would have been funny if it didn't have such disturbing ramifications. Watching this host of adults, supposedly intelligent, autonomous beings, behaving like remotely controlled mindless automatons: all simultaneously standing up, sitting down, now to their knees but resting their bums, now forward on their knees, back on their feet, knees again, now line up, close your eyes, eat a wafer, back to your knees; and throughout, chanting, chanting, all in unison, monotonal, expressionless, zombie-like. Heads nodding, Pavlovian ingrained involuntary response whenever they or someone else said the word 'Jesus'.

She had seen it week-in, week-out as she grew up and thought little of it other than 'I'm bored' and 'This means nothing to me'. Coming back to it as an adult, however, it was just scary. She was looking at mind control. Human beings reduced to puppets by sheer indoctrination, force of habit and a kind of cultural inertia: we do this because we've always done it. But as Voltaire put it: once you can get people to believe absurdities, you can get them to perform atrocities. Once people have allowed themselves to become puppets, never questioning what they find themselves doing or why they are doing it, then they have arrived at a dangerously negligent level of intellectual abdication, and there are some very evil people poised to exploit that.

Thus she was thoroughly fed up hearing apologists saying, 'It's not religion that makes them do it.'

Wearing her two-piece mobile gazebo, she had

51

personally overheard Akim Hasan, the would-be jihadi, propose bombing a nightclub on the grounds that 'the whores dancing around almost naked are an affront to God and deserve to burn in Hell'. Another, Falik Souf, had opined that any Muslims who died in the same planned attack would also deserve their fate because they had forfeited the right to call themselves such, having insulted Islam by partaking in this western decadence. And for an encore, between them they assured each other that any 'good' Muslims who happened to find themselves passing the building at the wrong time and thus end up as collateral damage ought to be grateful, because they would go directly to heaven, involuntary martyrs but martyrs nonetheless.

Any candidates for what could lie at the root of this kind of mangled logic or homicidally censorious morality? Listening to death metal, perhaps? Playing too many nasty computer games? The decline of the nuclear family? Choice of aftershave? Or how about an utterly irrational belief system that prizes faith over logic, dogma over compassion and slavish obedience over rational self-analysis?

Just a thought.

She: Angelique de Xavia. Ugandan Asian Glaswegian Poliswuman, lately operating out of Paris as part of an anti-terrorism task force for which she had been head-hunted by its chief, Commander Gilles Dougnac. She had just what he needed: intelligence training, Interpol contacts, several languages, a varied and resourceful facility for inflicting physical harm using both ballistic and manual methods, plus hard, wet, in-the-field

52

experience, having identified and taken down no less a figure than the notorious Black Spirit.

Dougnac had impressed upon her the importance of all of these attributes, as well as the depth of his professional appreciation for them, which helped him sell her the job. His remarks had been in contrast to the attitude of her superiors back home, whose appreciation of her efforts at Dubh Ardrain was rather disproportionately (in her opinion) tempered by the fact that, as well as accounting for the Black Spirit and his team of mercenaries, she had also accounted for a couple of hundred million euros'worth of hydroelectric power station. Thus Dougnac's job hadn't actually required much selling.

Retrospectively, she had come to understand that a further reason for his stressing what qualities she could bring to the job was in order to politely de-emphasise a more sensitive but even more crucial attribute: being that she looked a lot like the people she'd be up against. *La petite bombe noire*, as she became nicknamed: a formidable agent of infiltration, even without a burqa.

It was alarming how well suited she had been to her undercover work; alarming if, in moments of harsh self-analysis, unsurprising. The semblance of fitting in where she didn't really belong was a skill Angelique had lived by all her life; a social subterfuge mastered before she could read or write. She was born in Glasgow but her parents (and, strictly speaking, her older brother James) were Ugandan Asian refugees. She had been 'the wee darkie lassie with the funny name' at west-of-Scotland Catholic schools where having a granny from Ireland was the closest anyone had previously got to ethnic diversity. And she had been a woman—a dark-skinned woman—in the phalocentric, white-bread world of the Glesca Polis. Aye, she could write the manual for pretending to assimilate within a hostile culture.

This skill also had its inverse, however: she didn't fit in where she was meant to belong—her school, her town, her church—so she had found herself trying to belong where she didn't fit in. Hence the wee darkie lassie from the Catholic school had spent her teenage Saturdays watching Rangers, in the ranks of a support who had traditionally revelled in a triumphalist white Protestant supremacism. It had started via the logic that 'my enemy's enemy is my friend': everybody who hates me hates Rangers too. But in time she had begun to ask herself whether she merely found a paradoxical comfort in defiant isolation: that she never felt quite right *unless* she was somehow apart from the crowd.

All of which had ultimately led to one final relevant attribute: relationship status single. Dougnac hadn't made a big issue of asking, but checked cursorily, almost as an afterthought; or worse, merely confirming a foregone safe assumption. Foregone, safe, but it had to be checked. It carried with it a lot of assurances; for him, not her. Nobody to get resentful over you not being there for days or weeks, disappearing at zero notice, too alienated to speak when you did show up; and that only if you weren't anyway legally bound not to discuss anything. Nobody to miss you and feel neglected. Nobody to worry that you were in danger. Nobody for you to be thinking of enough to alter your priorities when the lead started flying.

In her professional life, the closest she had come to something she could call love had been with a bank robber, a fugitive from both the law and the underworld by the name of Zal Innez. If there was a starker example of seeking a sense of belonging in a place where she didn't fit in, she'd love to hear it. That had been five years ago, however, and his words on the last night they'd spent together still cast a grave doubt on what she really thought she'd been doing.

'When I called you on the phone, did you figure: "I know, I'll hook up with a bank robber—me being a cop, that sounds like a stable long-term relationship in the making"? You knew it couldn't last—maybe that was the attraction.'

Maybe. Maybe. It sure felt like something more at the time, but even then, she couldn't trust herself to know if what she was feeling was real, or merely what he had skilfully manipulated her to feel. He had been, after all, a master of illusion, thus there was no way of knowing whether he had merely fooled her like he'd fooled everybody else; and worse, used her as a part of the illusion that had ultimately ended with his disappearance.

All of which left her very single and, having failed to belong where she didn't fit in, singularly suited once again to fake fitting in where she didn't belong. She'd spent so much of her life concealing who she was that, in hiding her face too, the niqab was the only thing required to complete the effect: the ideal costume for Angel X, the superhero whose power lay in having no identity.

*　　　　*　　　　*

55

They had been tracking the cell—or constituent parts of it, at least—for almost a year. This had involved operations in Marseille, Toulon, Milan and Barcelona, as well as some vital intel from Islamabad, where several disparate members had been brought together for hardline indoctrination before being relocated to Paris. Angelique, in her Millet's camouflage outfit, had been able to plant a variety of surveillance devices in their meeting places and even their clothes and belongings. She liked to think she had more to offer undercover work than sneaking around under a shapeless black sheet, but there would have been little for her to glean under any other pretence. There were no double-agent or middle-woman roles for her to play. These people didn't talk to women; not about their jihad anyway. She knew this because she *had* been able to engage in talking to certain of their wives. She had a name and a cover story, and had even been seen without burqa or niqab while performing ablutions in the women's wudu.

Part of her cover was that she spoke no Arabic, which allowed her to eavesdrop on conversations intended to exclude her, another situation assisted by one's facial expressions being conveniently obscured. Her conclusion was that these women were either the coldest and most loyal sleeper-ops the Islamists could possibly pray to recruit, or they knew nothing whatsoever about what their husbands—the fathers of their children—were neck-deep in. Angelique's observations strongly suggested she lean towards the latter. The most common sentiment she heard made about any of the suspects by their wives was that they seemed preoccupied to the point of disinterest—in them or their kids—and consequently paid them precious little attention, far less confided about anything important. There was no chance, for instance, of them discussing with their women the issue of household finance, otherwise their spouses might be asking why anyone who lived in a third-floor apartment might require to purchase two dozen gallons of fertiliser.

Dougnac's unit had known about the nightclub plan for some time, but they were waiting—like the cell itself—for the arrival from Pakistan of their mentor and facilitator, Abu Syed, who was required to sanction their scheme and to procure them a source of reliable detonators. The unit wanted to trace this source and monitor subsequent traffic, and it desperately wanted Syed tied to something they could nail him for.

The detonators arrived from Marseille the day before Syed flew in from Islamabad. Members of the group were followed and filmed performing a test detonation at a disused quarry, under instruction from Syed, while Angelique's bugs picked up coded discussion, to which Syed was party, of the target location and the planned date of the attack, two months hence. As soon as they had that, Dougnac was satisfied it was enough, and ordered simultaneous overnight raids at seven addresses. These had taken place at zero four hundred hours that morning.

The day before a raid, you keep tabs on the suspects to make damn sure they're where you're expecting them to be. Then you go in under darkness, during the early hours, late enough for nobody to still be up, too early for the sharpest riser to be awake. Nobody gets time to react; they usually don't even get time to dress. It's standard procedure. Everybody knows this. But the reason it's so effective is that the subjects never realise they're under surveillance, far less have any inkling that you're planning a swoop.

So how come she ended up at the mosque six hours later, trying to lock down three fugitives before they could turn their place of worship into a televised siege and thus a worldwide propaganda coup?

There were explanations she couldn't begin to contemplate, such as a leak from the police side. There was plain old bad luck: one of the subjects gets up for a pee in the night and happens to look out the window just in time to glimpse a team of armed cops taking position outside; or an unrelated siren (as was heard in earshot of two of the locations) spooks a light sleeper enough for him to check the window, giving the same result: a vital head start to phone his buddies and bail. But most relevant, in Angelique's opinion, was the arrival of Syed.

The cell had given no indication they had any suspicion of surveillance, but he may have told them to assume it. A lucky break might have given one suspect a brief warning, but it would have made little difference to a simultaneously staged snatch operation unless they had a procedure in place. Upon Syed's advice, as well as an emergency escape protocol, they may even have had a look-out roster. One of the detainees' mobile phones showed multiple messages having been sent approximately two minutes before the squads went in, while the personnel were still mustering at points around the various locations. Despite the snatch teams operating in synchronous coordination, with stealth and maximum rapidity, three suspects managed to escape from two of the seven addresses: Akim Hasan, his brother-in-law Jafir Khan, and Falik Souf.

Cars were stopped, streets were blocked off, buildings evacuated, but to no avail. Sometimes two minutes is enough to vanish, especially in a city like Paris.

It was embarrassing, but not a disaster. The police had the explosives, they had averted a horrific bombing and, best of all, they had Syed. No need to panic: they'd get the others in time.

And lo, only a few hours later, the tip came in, from the Imam at the mosque on Boulevard Aristide Bruant. They had turned up there around nine a.m. seeking sanctuary, which the Imam was obliged to offer. He had intended to talk them into giving themselves up, but soon realised that wasn't going to happen: they were all armed, and he heard talk of something they had previously hidden in the building. He didn't know what it was, but assumed the worst. The mosque was already busy with classes in the madrassa and morning prayer was due to commence. He wanted to evacuate the place, but realised this could lead to an escalation of the situation, and he didn't want these fugitives in an even greater state of panic. Dougnac concurred.

The Imam was told to carry on as normal, while Angelique was told to suit up. That used to just mean Kevlar. At least on this occasion, the need for freedom of movement dictated the jilbab and niqab combo rather than the burqa.

Her job was to get in as close as possible to the suspects and monitor the situation, providing an ongoing sit-rep while the rest of the unit remained in a holding pattern out of direct sight of the building. The plan was to wait until morning prayers were over and the madrassa classes dispersed, then further personnel would infiltrate, dressed as worshippers. They would take the trio down without anybody hearing a thing, and, with any luck, without any of the faithful even knowing it had happened.

The plan was . . .

How many debriefs—formal or even inside one's own head—began with those three words? None describing ops that went right. No need to recall your intentions if your actions delivered the desired result.

Angelique had only located the first of the fugitives when the plan went out the window; or, more accurately, off the minaret.

The initial indication that the situation might be 'fluid', as the code euphemism had it, came through her earpiece with the report that several outside broadcast vans had begun appearing at speed, each pulling up on the nearside of the boulevard, and the optimistic possibility of it being a coincidence evaporated with the confirmation that the vehicles bore the logos of different networks: TF1, France2, BBC, CNN.

So it wasn't only the police who got a tip-off, but the broadcasters hadn't got theirs from the Imam. He was a Sufist, a vocal critic of the Islamists who had acted to prevent the likes of Hizb ut-Tahrir and Jamaat-e-Islami from poking their insidious tentacles directly into his mosque. To the

Wahhabist Nazis, he was *kuffar*, not a real Muslim; indeed, anybody who disagreed with them was not a real Muslim. He had stood in their way and this was where they got their revenge. They knew he would play it straight, call in the cops, and they knew a mosque synonymous with moderate, western-friendly Islam would be the ideal setting for what they had in mind. The Imam, Angelique rapidly understood, had merely been used to set in motion something that she was already caught up in and unable to stop.

The networks had been tipped off by the fugitives themselves. Angelique deduced this because it was only once all the TV cameras were set up that one of them leaned out of the minaret and began firing a pistol into the pavement directly in front of the news crews. It's doubtful any of the cameramen would have been fast enough to get that shot, but they'd have missed none of what happened next.

A gunman is firing shots from a tower on to the concrete of a Parisian thoroughfare. Suddenly you've got armed police in body armour: no choice but to become highly visible and ward any passers-by clear of the field. Suddenly you've got people inside the mosque hearing gunshots, seeing armed cops swarming around their place of worship. Suddenly you have panic, chaos, a global television audience and invaluable propaganda.

The situation had rapidly become very clear to her. This was more than an evacuation protocol for aborting the cell. Syed had improvised a full-scale back-up plan, and it was in play.

Unfortunately for Akim Hasan, however, the problem with improvisation is that it can be great in principle but lacking in the finer details. He didn't look like he knew what to be doing with himself when the cops showed up. He'd just been hanging around the passageway outside the madrassa, then stood and watched everybody pile past. It was only when the armed cop had appeared that he took a hostage, despite Angelique hovering around and deliberately presenting herself to him for what she predicted would soon be his only option. It was like the magician getting the volunteer from the audience to pick precisely the card he needed him to. It was called a force. She'd learned this only a few miles from where she was now standing, but that was another story.

Angelique speaks into her mike: concealed, like her earpiece, beneath the niqab. It had been through this mike—via a short relay up the chain and back down again—that she had let the armed officer know it was she who was being used as a human shield.

'Hasan is neutralised, sir. A little too easily,' she adds. 'He was isolated. Something doesn't smell right. Do we have the others?'

Dougnac fails to respond. Instead she hears him fire off orders to officers outside the building, where it was chaos. There were people streaming out on to the boulevard, perfect cover for the remaining fugitives to slip away in the melee. He was trying to get them contained but beyond any potential field of fire; an undertaking involving armed policemen herding Muslims around the exterior of their place of worship in front of a host of television cameras. If she had to come up with a name for it, she'd have called it Operation Lose-Lose.

'Do we have the others, sir?' she repeats.

Dougnac still doesn't reply, but she garners some information from the rest of the chatter in her earpiece. They have one fugitive pinned down in the minaret; believed to be Jafir Khan. He had fired the shots but didn't get down quickly enough. Now he was hemmed in, trapped in the tower but armed and with the benefit of an elevated position somewhere on the spiral staircase.

'Nobody goes into the tower,' Dougnac orders. 'I don't want to give them a siege, but I want even less to give Syed the propaganda coup of shooting dead a Muslim in a mosque. Where's Souf? Do we have him yet? What about the Imam?'

'Stay with the prisoner,' Angelique tells Dumarque, then ventures through the far archway and into the madrassa.

The iPod muezzin is still calling from on high, the schoolyard hubbub of over-wrought departing voices quietening as the mosque is cleared. That's when she makes out the voices from somewhere above, the gallery level overlooking the haram. She hears a woman almost shouting through tears, indignation keeping her just the right side of hysterical; and a man's voice, lower, appealing but insistent, the sound of restraint. Angelique runs through the madrassa, between tables bearing hurriedly abandoned Korans and open jotters.

'First floor,' she reports. 'Possibly another hostage.'

She climbs the stairs delicately, picking out her steps on light and quiet feet. She's got her pistol holstered at her thigh, but she leaves it there for now. As before, the harmless irrelevance the niqab cloaks her in will offer the best protection while she assesses the situation. Her stealth is a wasted effort: the voices get louder as she draws nearer, oblivious to her approach and, it seems, to even the police incursion taking place all around. Angelique reaches the head of the stairs and sees them standing in the passage: it's not Souf, but the Imam, his hands imploringly gripping the arms of a woman. Her back is to Angelique, so she can't see her face, but she can see the small child clinging on to the woman's leg.

The woman is trying to push past, the Imam barring her progress, appealing for her to turn back. They are arguing in Arabic. The child is crying, distressed by the raised voices and the awful anxiety that is thick in the air between the adults.

The Imam notices Angelique.

'What are you doing?' he demands. 'Get away, get out of here. It is not safe. Go to the police.'

Angelique removes the niqab. 'I am the police, sir,' she says.

The woman turns to look at Angelique, which is when she sees that it is Souf's wife, Raziya. Her cheeks are tear-streaked, her expression gaunt and anguished. She looks back uncomprehendingly for a moment, her mind taking a while, in its shaken state, to put together why she knows Angelique's face. Angelique braces herself for anger; instead Raziya appeals to her.

'You must help me. Help me stop him.'

Angelique looks to the Imam.

'What is Souf's wife doing here?' she asks, for the benefit of her superiors listening in as well as the Imam.

She hears two almost simultaneous replies, one

64

in her earpiece and one from Raziya. The police hadn't held her or Hasan's wife after the raid, but had assigned officers to follow them in case their movements led to the suspects. Raziya had taken her son and left, well before the Imam made his call; thus when her destination became apparent, nobody had considered it particularly significant. She prayed there every day, and took her son to the madrassa several times a week.

She had come seeking the Imam, not knowing her husband was in the building, but it was her husband she needed to talk to him about.

'He made a videotape,' Raziya tells Angelique.

Oh shit.

'Can anyone confirm Souf left a videotape?' she asks.

'Officers are still collating the evidence we found at this morning's locations,' Dougnac tells her.

'I saw them,' Raziya insists. 'Two nights ago. They locked me out of the room when I . . . I told myself it wasn't . . . then this morning . . . please, we must stop him.'

'We already have, Raziya. We got the explosives,' Angelique assures her. 'Please tell me we got the explosives,' she adds, under her breath.

'We did,' replies Dougnac, 'and the detonators, but we cannot assume we got it all.'

Fuck.

'Where is he?' Angelique asks the Imam. He closes his lips tight. He's taking a beat to think what's best to say, but he's already confirmed to Angelique that he knows. He just doesn't want to tell Raziya.

'Imam, where is he?'

'I will tell you. But only if Raziya takes herself and young Saadiq to safety.'

'I'm not leaving,' Raziya insists. 'I don't want him getting killed. I can make him listen to me, if only I can speak to him.'

Angelique looks at the pair of them, decides Raziya offers a better chance of talking him into giving himself up than the Imam does.

'She's right,' she tells the Imam. 'You take the child.'

'No,' Raziya says, almost a scream. 'He needs to see his son, needs to see what is better than this . . . this *madness*.'

Angelique knows she's got a point, but the feel of the handgun strapped to her leg tells her this is just not happening.

'Imam, take the boy, and tell us where Souf is.'

Raziya looks at the three other people in the passageway, then down to the haram, where she can see more police moving in. She steps away from the boy, taking his hand from her leg and giving it to the Imam.

The Imam puts an arm reassuringly around the boy's shoulders and nods to the door of his office, just along the passage.

'There is a maqsura, a hidden sanctuary, adjacent to my personal chambers.'

'Get the boy out of here.'

'God be with you,' he says.

Aye, right.

Raziya is already running, tugging at Angelique's robes as she passes.

'Quickly, please, before the other police get up here. The ones with the guns.'

Angelique turns and accelerates, as fast as her

encumbering costume will allow, stopping Raziya's hand before she can throw open the door.

'Careful,' she warns. 'Slow. Don't get yourself shot. Call to him first.'

Angelique opens the door slightly and Raziya calls out her husband's name.

'Falik, please, it is Raziya.'

There is no response. Angelique nudges the door ajar inch by inch and peers inside. The chamber is empty. Raziya all but barges past her into the room, calling out 'Falik, please,' her voice broken with sobs.

They both stand just inside the Imam's chamber. It is a long and spacious room, running maybe a third of the length of the first-floor passageway. There are bookshelves, cupboards, two desks, one bearing a leather-bound open copy of the Koran, the other a flat-panel monitor and a keyboard. Two arched windows afford a view of the adjacent tree-lined boulevard. The two women's views are drawn not to the far wall of the room or the one abutting the passage, but to the wall in between. There is an ornate wooden screen, ostensibly ornamental upon first look, but closer inspection reveals the bottoms of two small wheels at its base. This is the maqsura.

'Falik, there are police coming, they have many guns. Please, don't let them hurt you. Saadiq is here too. He's with the Imam. He needs you.'

At last, this elicits a response from the hidden chamber behind the intricately carved screen.

'Why did you bring Saadiq? Get him away from this place.'

In her earpiece, Angelique can hear that her colleagues are taking position in the passage outside. Very softly, she tells Dougnac to keep them back from the Imam's office.

'I brought him here to show you what is important, to waken you from this madness.'

While Raziya talks, Angelique approaches the screen. She can't see light or movement behind it. There is white-painted wood behind the carvery work. Raziya makes to move forward, but Angelique holds up a palm to tell her to stay where she is. This is partly so that she doesn't get desperate and rush the screen, but mostly so that, with Angelique's back to her, Raziya doesn't see her unholster her pistol from her calf and conceal it in the folds of her robe.

'I am doing God's work. *Wajib*, Raziya. The only madness is in the minds of those who would oppose His will.'

His voice is edgy, breathless, wavering. Sounds like his heart-rate is off the scale. It is, worryingly, a little more anxious than she'd expect in somebody merely digging in for a stand-off.

'Falik, listen to your wife,' Angelique says, watching the screen for a response to this new voice. She sees nothing, but can hear movement behind the panel. 'There are armed police all over the building; they are right outside the Imam's chamber. You can't hide from them and you can't defeat them. Think of your family. Do you want your son to see you carried out of here? Think about Saadiq, think about the wife you have, look what she's been prepared to do for you, how brave, how much she loves you.'

'Please, Falik,' his wife implores.

'I'm a police officer,' Angelique continues. 'I can lead you out of here, in my custody, and

68

nobody has to get shot. Please, Falik. You can't serve God or your family if you're dead.'

'You know nothing of my family,' he retorts, shouting, over-wrought, almost hyperventilating. 'And you know nothing of serving God. But I will teach you.'

Angelique doesn't like the sound of this and feels herself change her stance reflexively. She takes a step back, legs slightly apart, and brings the gun up to eye-level, held in both hands.

As she does so, the screen rolls back.

'Nooooo!'

Raziya half moans, half screams, horror and despair, her worst fear, her greatest dread, the death of her hope.

Falik stands in the alcove, a backpack strapped in front of him. There are cables emerging from it on either side, snaking their way to his hands, which are held apart at arm's length, each curled around a thick, solid, metal cylinder. He is bathed in sweat, like the maqsura was concealing a sauna. His breath is rapid, his eyes bulging, as though he's in some state of hyperconsciousness. He stares at Angelique, moving his hands slightly closer, then stopping them. He's letting her know: the cylinders touch, he completes the circuit.

Dougnac's voice is in her ear, asking what's going on.

'He's got a bomb,' she reports, as steadily as she can. 'Pull everybody back. Get everyone out, *now.*'

Raziya drops to her knees, wailing, hands clasped.

'If you've got the shot,' Dougnac orders, 'take him. Immediately.'

Raziya is pleading with both of them, her words

69

now a babble, distilling finally to one: 'Please. Please. Please.'

Angelique has the shot, but she disobeys. She's not going to shoot this man in front of his pleading wife until she absolutely needs to.

'Put down the detonators, Falik.'

'Take him, now,' Dougnac commands.

'You're not going to blow up a mosque,' she states, her appeal a simultaneous explanation to Dougnac: she still thinks she can talk him down.

'What about a mosque with twenty crusader cops inside, who have "violated its sanctity",' Dougnac responds. 'He blows the place up and the militants can make up any lie they like. Say *we* blew the place up. Anything.'

Fuck.

'Think of your son, Falik. Saadiq is downstairs. Do you want to kill him, and your wife?'

With the mention of Saadiq, Raziya collapses like an abandoned puppet, bawling in anguish. Souf looks to her, then to Angelique.

'Don't cry,' he says, swallowing, his voice a hoarse dry whisper. 'You will not lose Saadiq. In mere moments we will all be together. In paradise.'

Angelique breathes in, remembers the other things she's heard him say, knows he believes it. He brings his hands together. She shoots him, drilling four shots through his brain in less than a second.

Raziya screams.

'Suspect down,' Angelique reports, getting the words out before her voice chokes.

Souf's body slumps into a sitting position, his back against a chair inside the concealed sanctuary. His hands drop limp by his sides, and out of each rolls a twelve-volt torch battery. The cables dangle loose now from the backpack. To the right of the body Angelique can see a standard lamp, a short, frayed stump of flex jutting from the bottom, and a discarded plug-top close by on the floor.

His wife runs towards him. Angelique can't bring herself to intervene. Raziya kneels down and throws herself around her dead husband, wailing in grief.

The plan was . . .

Syed had got his martyr, got a Muslim killed in a mosque by the *kuffar*. What an obscene waste of a human life. But, let's all say it together, religion wasn't to blame.

Angelique sees armed cops approach the doorway. She waves them back. 'There's no bomb,' she reports. 'There was never a bomb.' Her vision blurs, which she suddenly realises is because she is crying.

Raziya looks up at her from the body. There is no anger, no accusation, just a lost, hopeless incomprehension. Asking why. Why did this happen to her, to her husband. To her son.

Angelique looks down at the gun she is holding, looks long at the device and the hand that wielded it. A thought briefly strays into her head that she does not want to admit to. It passes. She breathes out, flips the safety, holsters the weapon. Then she looks at the mother on the floor.

A mother: that's what a woman is supposed to be. Women are meant to create and nurture lives. All she's done is end them. Saved a few, but the goalkeeper's contribution never registers in the annals, does it?

71

The siege, such as it is, ends quickly after that, or maybe it just seems that way in Angelique's numbed condition. Her actions having popped the Parisian cherry on in-mosque shootings, Dougnac has no further reason for entertaining Jafir Khan in a stand-off. The minaret gets stormed and he is taken down by two shots below the waist. He'll live.

There are paramedics hurrying into the courtyard as Angelique exits the mosque. She sees the Imam standing behind a cordon in the company of several uniformed officers, his arm still around the shoulders of the boy. Angelique averts her eyes before the Imam can catch them, and in greater fear of meeting those of the child. She is aware they don't know yet. She feels as if she could convey everything with just a glance. It's only one of so many reasons she can't look either of them in the face.

She slips away through the gathering crowds, past the cameras, to a phalanx of police vans around the corner. She takes off the jilbab and squats down beside the rear left wheel of one of the vehicles. Somebody hands her a bottle of water and she drinks from it, staring ahead, seeing nothing. She sits there motionless with the open bottle in one hand, the lid in her other, like she's in stasis, a machine broken down.

'How you doing?' asks a voice beside her. She looks up and sees the silver-haired figure of Gilles Dougnac.

At last, she is reanimated. She has another drink from the bottle and climbs back to her feet. She takes her time swallowing the water, all the time looking him in the eye as she does so. She wipes her mouth.

'I quit, is how I'm doing.'

GLITTERBALL SHARDS (II)

The magician's hands suddenly spasm as he grips the pack. The cards explode from the collapsing cradle of his fingers, spraying, spinning, fluttering about the stage like crisp autumn leaves stirred from the gutter by a sudden gust. It is not a flourish, but a fumble, a moment of startlement. A trick derailed, an unscripted incompetence. Some members of the audience gasp, others fail to stifle giggles. The muted laughter is horrible: a cringing combination of being embarrassed on the faltering magician's behalf and being embarrassed by being present at such a tawdry spectacle. But can he recover, that's the question? Does he have an out?

PORTRAIT OF A LOST SELF

It's three days later. She's in a cottage six miles outside Chartres, lying in the bath, wishing she could remain there in a warm and bubble-covered state of suspended animation.

She had to get out of the city two days back merely to get some sleep. She remembers forcing herself to turn off the TV, with its endless looping of the same images, feeding the news cycle that was spinning inside her own head. Footage of the mosque, police gathering, worshippers running out, being corralled at what unavoidably looked like gunpoint, the sounds of shots, the wail of ambulances, the face of Raziya Souf, screaming her loss. The cycle evolved to include the first gatherings of masked youths in the suburbs and the earliest pronouncements of exuberant outrage from the Middle East. By dusk, there had been cries of vengeance: nauseatingly predictable demands for the deaths of those responsible for this affront to Islam and, in particular, the 'murderer' of Falik Souf.

'He must pay for violating this holy place,' raged some bearded twat in Saudi Arabia. 'We demand his name and we demand his head.'

Time was when Angelique's response would have been a wry smile at how much this windbag would enjoy meeting 'him' one-on-one to discuss matters, but not any more. All she could see was a wave of hate, pulsing out from an epicentre at which she stood over the body of the man who had fooled her into making him a martyr. The shockwave had already made it halfway round the world and back again. There were 'spontaneous' demonstrations being orchestrated throughout the more militant enclaves of the Islamic world; a good day for anyone with a stockpile of the *Tricolore* to sell. Buy your French flags here, free matches and lighter fuel with every purchase. And by nightfall in Paris, there was rioting on the streets: live footage of cars on fire, stones and petrol bombs arcing towards police lines, all against a soundtrack of sirens.

When she turned off the TV, she could still hear the same sirens from less than a mile outside her window. That was when she got in the car and left for the rural bolthole she co-owned.

'Take some time off,' Dougnac had said. 'As much as you need.'

'I reckon about twenty-five years should cover it,' she had replied. 'I'm serious, sir. I can't do this any more. It's over.'

'This will all feel different in a week, maybe less,' he assured her.

Three days on, what worried her most was that he would turn out to be right.

It had happened plenty of times before. She hadn't actually announced her intention to quit before, but she'd been through the feeling that she couldn't go on, couldn't recover from what she had just been through sufficiently to pick herself up and head back into the fray. Yet, every time, she had: no matter what she had witnessed, no matter what she had been required to do, no matter the danger or pain she had endured. The scar tissue, she knew, would soon form over what happened at the mosque like it had formed over so many previous wounds. Each time she went through something like this, she emerged just a little bit tougher, which meant that she was able to feel just a little bit less. In this instance, perhaps a lot less, and that was why she feared Dougnac being right. In a few days, things *would* feel different, and in time she would be able to go back. But what would be left of her inside?

All those times before, when she had cried herself dry and then returned to the unit, it wasn't because she was healed: it was because she had nowhere else to go, nothing else to do, and certainly no one else to be with.

The job never loves you back. It chews you up and spits out a husk.

There was a painting she kept in a cupboard in the cottage, still protected and obscured by its swaddling of cellophane bubble wrap. It was painted by Zal Innez, a coded invitation to a rendezvous in another country (not to mention another legal jurisdiction) from the one where their paths, having been thrown into collision, would ever after have to remain parallel, never able to touch. There were no expectations or conditions attached, just a secret where and when, encoded in oils, allowing the possibility, despite their distant and secretive lives, of simply seeing each other again. The when was her birthday, a year on from the day they met. The where was the Musée d'Orsay.

When the day came, she was three time zones away, sitting in a damp wooden hut in Lebanon, surveilling a farmhouse on what had turned out to be bad intel, unless what her binoculars had shown her was actually the foundation of Al Qaeda's first jihadi-goat training camp. The job saved her making a decision about whether to show up, and taught her what her decision always ought to have been anyway.

The closer the date had approached, the less sure she had become of whether she intended to appear, and she had barely dared contemplate the question of whether she believed Zal would.

When she was being most rational and least insecure, she was at least able to convince herself that what had happened between them was real. Zal, for all his abilities to deceive, had opened up to her, rendered himself vulnerable in ways that she refused to believe any man could fake, or would possibly wish to. He had exposed the scars on his soul for her to read and understand like she had read the tattoos on his skin. But it was that same tenderness, that same vulnerability, that told her how, even if she wanted to see him again, for his sake she ought not to.

If you love somebody, set them free, wasn't that what they said?

She had no idea how much anyone else might have known about their relationship. DCS Jock Shaw had remained unreadable on the matter, both at the time and in the aftermath. He knew she had been secretly meeting with Zal even while half the Glesca Polis were out looking for him, but as the intelligence deriving from this had proven expedient towards nailing a far bigger target, Shaw hadn't probed too deeply into what these meetings had involved or might imply. However, what he suspected about it—and what he might even have secretly discovered about it—was an unknown variable in an already very unstable equation. All of which was to say nothing of the fact that the Royal Scottish & Great Northern Bank remained eight hundred thousand pounds light and denied even the satisfaction of seeing a suspect in custody.

After everything Zal had endured in his life, through loss, through prison, through tragedy, through pain and fear; and the labours he had undertaken to extricate himself from the dangers of such things being repeated, she couldn't see herself as anything but a risk, and one that she had no right to expose him to. What could she give him anyway, in exchange for that risk? The occasional off-chance of some time in the company of a knackered, stressed and increasingly alienated policewoman? It wasn't like she was going to quit her job and run off with a fugitive, was it?

You knew it couldn't last—maybe that was the attraction.

It was a week after her birthday when she got back from Lebanon. Some small part of her succumbed to wondering whether he might yet show up, having found out where she lived; or whether he might just appear, with typically magical showmanship, were she to go along to the museum several days late. But these were just the daft, girlish dreams that were hard to give up when reality required a difficult, adult decision. In this case that decision was to let it go: that it was best for both of them—and for Zal in particular—if she just let him think she didn't want to see him again. That, in a way, would be the most loving thing she could do for him: let him forget about her, let him escape the last threat that might yet see him tracked down and lose everything again.

The painting remained wrapped up, never

79

looked at it in all the years since its invitation expired, but she had not been able to bring herself to throw it out either. She had never considered hanging it, simply because she was too self-conscious to have a painting of herself staring back at her, but as the years had passed, she became more and more reluctant to even take it out from the back of the cupboard and have a peek at the thing. She was afraid to look at it, like it possessed an inverse version of the Dorian Gray effect: a face that would provide a disturbing contrast to the one that tended to pitch up on the mirror. The woman in the painting was not just five years younger, but seen through the eyes of Zal Innez. It was a portrait of a woman she had once wished she could be: a happy woman, a hopeful woman, a woman who might be falling in love. Now, what worried her wasn't just the fear that she'd never be that woman, but that she was no longer the woman who once wished to be.

'This too will pass.' That was one of Zal's tattoos: a sentiment intended to give hope in times of hardship and sound a warning in times of joy.

This will all feel different in a week, maybe less, was how Dougnac had unknowingly echoed the same point.

The corollary to both, however, was the favourite of an elderly but sprightly neighbour when Angelique was growing up in Glasgow: 'You're a lang time deid.'

Enough.

Sorry, Gilles. This time she wasn't coming back.

She hears the phone ring, ignores it. Probably for Mireille, with whom she co-owns the cottage. Angelique has given the number to very few people, and none that she's prepared to speak to right now.

Her mobile rings, only moments after the landline gives up. She ignores that too. It rings twice more, each time diverting to the voicemail service. She decides to switch it off altogether so that she can enjoy her bath in silence. She gets up and wraps a towel around herself so as not to soak the carpet as she retrieves the phone from the coffee table in the living room. The bloody thing rings again just as she's reaching for it, and she can't help but see the caller's name:

Jock Shaw.

She answers, some automatic reflex kicking in even though it's been several years since she was answerable to him. Jock Shaw, for a long array of reasons, is the kind of guy you'd still take a call from if you were hanging from the window ledge of a burning building.

'DCS Shaw. It's been a long time. Sir,' she adds.

'Mademoiselle de Xavia. Playing hard to get.'

'I was in the bath,' she explains, simultaneously wondering why she feels the need to do so. 'What can I do for you, sir?'

'Ach, it's just one of those late-night drunken phone calls, hen, you know? Had a load of bevvies with some pals and got reminiscing about an old, eh, acquaintance of yours, not seen round these parts for a good few years. Led to your name popping up, so I thought I'd give you a wee bell just for old time's sake, you know?'

Angelique feels herself shudder and stiffen. Shaw sounds perfectly sober. Her mind becomes a zoetrope of possibilities, her pulse speeding up like the strobe light inside. Fear that her secret past is about to be exposed, fear that her beloved thief is in danger, the tantalising notion that she might see him again. So many flickering frames coalescing into one image: Zal.

She says nothing, a long nothing. At the other end, she hopes, it just sounds like she's not in the mood for banter. She swallows, clears her throat.

'What do you want, sir?'

'Heard you quit.'

'You heard right.'

'How many times is that, now?'

'Just the once, sir,' she responds calmly.

'Serendipitous.'

'How so?' she asks, then reads ahead. 'Oh, no, don't even begin—'

'Monsieur Dougnac reckons you just need a break. They say a change is as good as a rest.'

'I'm taking more than a rest. Dougnac's in for a disappointment.'

'I don't doubt it. But that doesn't change the fact that we've a need for your talents back here. Something has—'

'When I said I quit, I meant the whole shebang, sir.'

'Hmmm,' he says, with the infuriating certainty of clearly having something up his sleeve. 'You seen the news lately?'

'I've not turned on the TV for two days. If you've spoken to Dougnac, you'll know I *am* the news.'

'I was talking about the big story back in Blighty. Thought that might pique your interest in a wee trip home.'

'What is it? Are they abolishing the monarchy? I'd come back for that.'

'You heard of Darren McDade?'

'The wee prick that writes send-them-back and string-them-up stuff for one of the tabloids?'

'That's the chap. Disappeared a few days ago. Turns out he's filed his last column. Somebody killed him, and went to an awful lot of trouble about it.'

'Not really registering on my heart-rending tragedy metre,' she says. 'In fact, I think I just felt the world get more tolerant by the weight of approximately fifteen stone of arsehole.'

'Well, I'd hold off on the street party just yet. You got access to your email there?'

'Yes, got a laptop.'

'Good. Call me back on this number once you've checked out the links I've sent you.'

And with that, he hangs up before she can respond.

She boots up her laptop.

Forty minutes later, she calls back.

'This stuff is all over the internet, then, I take it?'

'Aye. McDade's newspaper and the law have been busy getting the files pulled from various websites since about two hours after the videos first appeared, but they pop up elsewhere just as quick, spreading like plooks on a teenager.'

'I believe the technical term is "viral", sir.'

'See, that's why we need young officers who are *au fait* with modern technology and hip to the groove, daddy-cool.'

'Aye, very good. So where do I come into your thinking? I'm assuming from the faux-Guantanamo clobber you reckon we're dealing with an Islamist group.'

83

'Well, my eldest suggested it could be somebody who really hates Slipknot. I had to ask him to explain that one.'

'Was that before or after you asked him to explain what he was doing streaming execution videos?'

'Well, it was him that alerted me to the situation, a good while before any of my fellow polis would have done, so I let him off.'

'Joking apart, what's the script? Anybody claimed responsibility for this?'

'After a fashion.'

'After a *fashion*? What the hell does that mean?'

'It means you're slipping in your old age, Angelique. I'm sending you a new link just now. Okay, we had to digitally enhance a detail of the video concerned, but time was you'd probably have spotted it with the naked eye.'

Shaw had a unique ability to flatter you and take the piss at the same time. It was probably one of the reasons he managed so effortlessly to command so much loyalty. She refreshes her mail client and clicks on the new link. This time it is to an encrypted police server and not some public web vault. Shaw dictates a username and passcode.

It's from the final video, the actual execution scene. Angelique feels a knot in her gut as soon as it flashes up. She was hoping not to have to watch any of it again, but this part was in a whole other domain of sickening. At least the sound is muted. The killer had bound McDade but didn't gag him. He wanted to hear his screams, his tears, his pleas for mercy.

Like all of the clips, the scene has been staged so that the killer never appears in shot, other than

84

the occasional hand and, in one instance, his back from the shoulders down.

The scene pauses, leaving McDade dangling, his tied legs frozen in mid-two-footed kick. A cursor arrow appears in the frame and tracks to the right of the gallows, where it proceeds to describe a square of dotted lines around an area of what upon first glance had appeared to be merely a section of black wall. Now highlighted, Angelique notices a slight difference in light and texture, as well as some tiny specks of white against the black, which had previously resembled nothing more conspicuous than splashes of paint. The cursor arrow changes to a magnifying glass and begins to zoom. As the focus draws in, the difference in texture becomes more pronounced until Angelique perceives that she is looking at a reflection. Before the dotted square has filled the frame, it has become clear that the highlighted detail is a mirror attached to the rear wall, and she feels her mouth open as she recognises what it is reflecting.

'You are fucking kidding me.'

Taking up most of the screen on her laptop now is a blurry but unmistakable close-up of the killer's head, staring back towards the camera, having taken up position quite deliberately to supply this signature detail. He is wearing a small black cowboy hat, his face covered by a black sheet. Upon this are drawn two rows of grid-like white teeth for a mouth and two white squares above for eyes, oversize black pupils in the centre, doubtless concealing view-holes.

Rank Bajin: cartoon villain of a 1960s Glasgow cartoon strip, partial subject of a much-loved statue in bronze on Woodlands Road, and erstwhile calling card of the rampagingly egotistical mass-murderer Simon Darcourt.

'Never did find a body, did they?' Shaw asks rhetorically.

Angelique feels her mouth going dry.

'Jesus Christ,' she breathes.

'Aye. They say *he* came back from the dead as well.'

BAR ACT (I)

Zal's hands fan the deck along the bar top, faces up, then flip them back one hundred and eighty degrees, as though he pulled a cord and they were strung like Venetian blinds. The red-and-white-checked backs of the cards form a continuous tessellation across a span of about twelve inches, broken only by the single all-white edge of the solitary card that is now facing the wrong way. The girl who chose the card blushes, gives the guy she's with an elbow in the ribs in response to some remark Zal never heard above the chatter around the bar. Zal places a finger on the inverse card and nudges it forward just enough to expose the corner and confirm that it is the nine of diamonds. The girl puts a fist in her mouth and bites on her knuckles to stem self-conscious laughter as raucous applause breaks out among the onlookers. The boyfriend shakes his head in amused befuddlement, then slaps a fifty-euro note down on the hardwood and tells everybody to order what they like. Amidst the dozen or so shouts for various drinks, there is a clamour for Zal to show the patrons something else, but he tells them, 'All in good time,' while Katarina and he get busy dealing with the orders.

It seems like everybody present takes up the boyfriend on his largesse, save a shy-looking young couple at a table on the raised area near the back—content to enjoy the show but maybe not comfortable accepting drinks from strangers—and a solitary male sticking to mineral water and paying the majority of his attention to his mobile phone.

Zal breaks the seal on a fresh bottle of dark rum and checks his watch while his wrist is turned to pour shots into three glasses he has lined up on the gantry. He's off in two hours, and there will be no hanging around after closing tonight. He's got a flight booked for the morning, though he's still not sure he's getting on it. He's told Franco, the bar owner, he'll be gone for three nights; he hasn't said anything about possibly not coming back. He doesn't wish to ponder what that might tell him about his own plans, expectations or even fears.

He's in a bar in the medieval part of Palma. It's just gone eleven, twenty-five hours short of marking one year to the day since he robbed the Royal Scottish & Great Northern Bank. One year to the day since he met Angelique de Xavia, and commenced this impossible but inescapable Escher painting of a relationship that had lasted only a couple of weeks but dominated and confused his thoughts thereafter.

After what went down in Glasgow, there was no question they both had to go their separate ways, nor that it was unwise for either of them to stay in touch with the other, but Zal had offered a means of marking the anniversary, just a way of leaving the door open.

Half the time, he found himself wishing he hadn't, that he'd just cut and run, done what he was best at. But in the lonely darkness of any night he found himself lying awake, he knew that this was something he couldn't let go. Not yet, at least. The idea, the possibility was not merely a comfort to cling to in an empty bed, but the closest thing he

88

had to a purpose in his life. However, that purpose, or deferred option upon a purpose, was due to expire. This Escher painting, this Möbius strip that seemed to offer two discrete realities, was about to unravel itself unalterably, leaving only one or the other. He had a choice to make, one further clouded by the knowledge that Angelique's own choice could render his moot. At least there was little chance of her forgetting the date: it was her birthday, after all.

It's not expired yet though, and maybe that's why he's booked the flight. He made a date, kind of. He conveyed an intention, and at this point, he's intending to keep it. He's going to follow this intention as far as he can. He's getting on that plane, he's going to Paris, and he's checking into that same Cubist hotel where they first spent the night together. But whether he's going to walk into the Musée d'Orsay, he still can't say. An impossible relationship, no less impossible one year on, so far as he's aware. But inescapable? To a man of Zal's experience, very few things could claim such status. Extricating himself was what he did best: the sacred art of leaving. It was the art of staying that he was yet to master.

It's been a quick year, a very quick year. A good year, a strange year, full of months that passed in a blur but days and mere moments that held him in their own stasis, seeming so distant from each other and from now that they felt as though they must have been decades apart.

He 'did' Europe, blended in with the tourists, stayed on the move. There was no way of knowing whether anyone was trying to trace him—the Estobals, Hannigan's mob, the RSGN, the cops—

and though he had taken care to cover his tracks, it felt safer to remain in this fluid state, this permanent transience. Travelling hopefully, never arriving: that's how this year had passed. After prison and everything he had been through in recent times, it was a convalescence, a sabbatical: time to recover and repair, to drift, to simply be. Time to contemplate, time to think. Just not enough time to reach any understanding.

He found himself in Palma late in the autumn, as the peak tourist season was winding down. He was tired of travelling and wanted somewhere to cool his heels for a while, that date in December looming not so far ahead. He chose Mallorca for a number of reasons, one of which had proven laughably moot: that he spoke Spanish. That said, it did throw some more dust over his footprints given that anyone trying to track him down would most likely be looking for an Anglophone. Again, this might have been a needless consideration, but better to be safe; thus another reason for the choice of the Balearics was that an island was a logistically tricky place from which to forcibly abduct someone. No protection ultimately if it was the law who came calling, but that was only one potential threat.

It was off-season, but the visiting population was still a reassuringly large—and reassuringly diverse—throng to be drifting within. There were a lot of ways to be nobody in particular on an island such as this, and he found old-town Palma to be just his speed.

Maybe it was passing easy time in so many bars throughout the year, maybe some vestigial prison-time daydream, but the idea of owning a little joint of his own had begun to take root. Perhaps not merely a bar, more like a cabaret club: somewhere with a dais just big enough for a baby grand and a microphone, somewhere for a little blues, a little soul, and—the growing kernel of a yearning he had to admit to himself—a little magic.

But shit, yeah, every guy in Folsom talked about owning a bar: that's what they'd do when that one big score came off. What they really meant was that they just wanted to *be* in a bar, and when you're banged up in Walla-Walla, who doesn't? But who the hell was planning robberies just so that he could end up working his ass off keeping a legitimate business concern afloat? Nobody lay awake at night and fantasised about arguing with city officials and licensing boards, or about negotiating with breweries and making kickbacks to suppliers. Thus Zal took the sensibly gradualist step of merely *working* in a bar for a while.

He didn't make a conscious decision to begin performing tricks. He'd been working at the Dracon Rojo for a couple of weeks when one night there was a middle-aged English couple playing cards at the bar, just passing slow time, long drinks and gin rummy, keeping score on the back of a beermat. As the place filled up, Zal noticed a young buck pestering them for a loan of their deck, with which he proceeded to try impressing a trio of adolescent German chicks. He was arrogant to the point of obnoxious; once he had the cards, he acted like the people he'd bummed them from no longer existed, the shy English couple left in a state of mild humiliation as they helplessly waited for the return of their cards. This pissed Zal off, but not as much as the fact that the German chicks were failing to shine the guy on. He didn't exactly have them eating out of his palm, but long as he was the only show in town, they were happy to let him entertain them—thus prolonging the older couple's discomfiture.

Some people feel naked or lost if they leave the house without their mobile phone or their watch; with Zal it's a deck of cards. He broke out a pack of his own and handed them to the couple as a replacement, but not before executing a couple of deliberately ostentatious shuffles in order to grab the attention of the German chicks and their performing monkey. As he intended, having passed his own deck to the gin-players, the girls insisted the current act hand Zal the first pack and leave the stage to the next performer.

'Let him show us a trick,' one of them said, and Zal was happy to oblige.

Monkey boy must have slipped away at some point over the next half hour, but Zal didn't notice him leave; nobody did, especially not the German chicks.

They came back the next night with some guys they had met, and insisted he show the new arrivals some of his stuff. Night after that, the three girls didn't show, but friends of their friends did, and so on.

Zal was enjoying himself. He worked the evenings and practiced by day; not necessarily stuff appropriate for the bar, just practice, practice, practice: sleights and subtleties, shuffles, false cuts, drops and palms, vanishes, transpositions, penetrations. There were a few other staples he worked on, definitely not for the bar, which made him ask himself how far he might really want to take this.

He practised even more than he used to in Folsom, and was only seldom reminded of how that felt as a means of whiling away those otherwise useless hours. It was easy to block the association, standing at a table in the courtyard outside his apartment's patio doors with the mild autumn sunshine on his back. No, it wasn't thoughts of prison that came flooding back as he felt the cards and coins between his fingers, but the bittersweet nostalgia of learning at his father's knee. He could lose himself in the repetition and concentration, barely aware of his surroundings, enough to become immersed in the most vivid recollections of that simplest, most innocent time. And though sometimes it threatened to be overpowering, and always it hurt just a little, he embraced it as a way to remember the best of his dad, the father as seen in the eyes of an entranced seven-year-old boy, before it all went alcohol-sour and his outright rejection of magic as a career was the best way Zal could think of hurting him back. The man he wouldn't forgive, no matter how much the old man tried to earn it; and now he couldn't forgive himself for withholding that blessing until it was too late.

A couple of months pleasantly passed. Business was good, tips were good, and Franco, the bar owner, was quietly pleased. Late autumn moved into the early stages of what passed around here for winter, but the Dracon Rojo was as busy some nights as might be expected in mid-August. Zal was aware, in some pragmatic vigilance substation in the back of his mind, that he was garnering attention, and that with this came considerations.

(He didn't always want to admit so much to himself as calling them 'risks'.) But he had to play the odds: it wasn't like he was executing a few neat card pulls then saying: 'For my next trick, I'll show you how the principles of misdirection can be applied to bank robbery!' Realistically, what was the chat? There's a barman at the Dracon Rojo who puts on a little magic show: that's it.

Nonetheless, people talked about the price of fame, and sometimes fame itself was the price. Jack Nicholson once remarked that when people said they wanted to be rich and famous, he'd suggest they try just being rich, and see how that works out for them. When it came to entertainment, to showbiz, you couldn't expect all the attention to remain directed at the stage once you had left it. All he was doing was performing tricks at the bar, and yet soon after he started it, all of a sudden everybody wanted to know him: people hung around to talk to him on his break, invited him out to clubs, parties or just back to their place when the bar was closed (though he never went). People would nod in the street: not just the nod of polite recognition and acknowledgement, but a knowing grin, like a shared secret. It made him a little nervous. There were, as stated above, considerations, but other considerations superseded them. Hard to tell Franco and the other bar staff that the golden goose was just going to quit laying, for one, but in the main, truth was he couldn't give up the juice he got from performing. How the hell did his old man end up an alcoholic when he had this to get high on every night?

Zal walks the short distance to his apartment

along the narrow streets, lanes and alleyways that look so peaceful. The tour guides just don't mention that they once literally ran with blood, when the Holy Inquisition oversaw the slaughter of every Moorish man, woman and child in the city. Slaughter on that scale, and yet the same buildings, the same cobblestones today look like something out of a fairytale. So much for ghosts.

He lets himself into the rented apartment on the ground floor of a four-storey block constructed when the average tourist carried a scimitar. The place has weathered the centuries pretty well, but at some point in its more recent history, maybe a hundred years back, a few iron pillars have been added for internal structural support. Well, Zal figures that's what they're for, but the one in his bedroom feels like it was made specifically for him to lean against with one hand, a beer in the other and unanswerable questions about the future streaming through his head. It's close on one in the morning. His flight leaves in nine hours. He hasn't packed a bag. He finds himself trying to interpret his own subconscious with regard to what this might suggest, comes up neutral: packing a bag will take all of five minutes. He pours himself a glass of water and heads for the bedroom. As he comes through the door, he sees his ticket and passport waiting for him on the nightstand. He'd normally not leave such items lying out in full view. Interpret that, he thinks: left visible because the trip means so much, or left vulnerable in case a burglar whips them and takes the decision out of his hands? Except, he's pretty sure he can't remember taking them out of the drawer before he—

He feels an explosion of pain across his back and nothing but white fills his vision before a second explosion sears the backs of his legs, which promptly buckle beneath him. As he drops to his knees, he catches a glimpse of something coming towards his head at speed, then all is white again.

Zal sprawls on the floor, the room spinning. He tries to roll on to his back, and as soon as he does so, he feels a blow across his midriff, then another to his face, this time partially blocked—at considerable pain—by a flailing arm. A further blow to the face causes his head to crack back against the tiled floor, making everything swim and his body go limp. He doesn't fully lose consciousness, but feels like a helpless ragdoll as he is hauled backwards against the iron pillar. His hands get pulled behind him and cuffed, then his feet also, leaving him slumped on his knees. His vision lurches like the drunken spinning that heralds the big spit, but he feels it steadying. He looks up, sees the figure of a man holding a nightstick.

Christ. It's the guy from the bar tonight, fucker who was just drinking mineral water and making calls on his mobile.

'You sitting comfortably there, my son?' he asks in a Cockney accent, filled with a cheeriness incongruously close to genuine warmth.

'I'm good, yeah,' Zal manages.

'Sorry about the tough love, but I didn't fink just asking politely would do the trick, know what I mean?'

'You could have given it a shot.'

'Maybe next time, eh?'

'Yeah. You wanna undo these cuffs and we can

96

start over?'

He dangles the keys between his fingers as though he's thinking about it, then slips them into a pocket in his trousers. 'P'raps not.'

Zal coughs, takes a breath, aftershocks of pain pulsing through him every few seconds.

'Who you working for?' he asks.

'Working for meself, mate. Each man is an island an' all that.'

'If you're done quoting Donne, I'll rephrase that. Who's paying you?'

'Ah, see, that all depends, dunnit? That venerable union of long-established and highly respected financial institutions, the Royal Scottish & Great Northern Bank, are apparently still feeling deeply aggrieved at the grave public offence and less than trivial financial injury inflicted upon them by some mysterious masked Yank with a flair for the flamboyant, ain't they? And they've put a value on the salving of said injury and offence to the tune of one 'undred grand.'

'So you're a bounty hunter. Do I call you Bobba or Greedo?'

'Greedy, more like. See, I ain't so bothered about dealing wiv the bank. They've just rather conveniently helped set a price floor for a little bit of 'aggling. Cause I reckon whatever they're offering, a certain Mr Bud 'annigan of Glasgow is gonna feel obliged to top.'

'Hannigan's in jail, or did his lawyer work his magic again?'

'Nah, he's all banged up, my son, but he still controls the purse strings, don't he, and I reckon it'll make those long winter nights pass that bit more pleasantly if he knows the reason he's there has been suitably chastised. Maybe he can keep re-watching said chastisement on a little portable DVD player what he's got stashed in his cell.'

'Sure, I've been there. Anything to pass the time. So how'd you find me?'

Greedy taps his nose and smiles. 'Trade secrets, my son.'

'And I take it another trick of the trade is gonna cover getting me off this island.'

'Nah, no big secret, mate. Just 'ope you don't get sea sick.'

'Figures,' Zal says resignedly.

Greedy grabs a drink from Zal's fridge and downs the can in one go.

'Cheers,' he says. 'Now, I'm just gonna go siphon the python and then we'll be on our way.'

'All that mineral water, good to flush the kidneys.'

'Quite right, squire.'

Zal watches him retreat and casts an eye towards where his passport and ticket sit, out of reach. This could sure make his mind up for him, but he's not going to let it. He can't guess how the guy tracked him down or whether his performing magic in the bar had anything to do with it, but he does know the asshole hasn't done his homework on conjurors. Coins and cards are your bread and butter, that's what his daddy taught him first, and as familiar objects, they can form the basis for tricks anywhere. But his daddy worked some big lounges in Vegas: you need more than close-up work when the audience gets larger, and it's not all done with smoke and mirrors.

Greedy emerges from the bathroom long

enough after the flush as to indicate he's washed his hands. Health and safety legislation must apply to even bounty hunters now. He returns to the bedroom with a canvas bag, from which he produces a clear glass bottle, looks about ten or fifteen mil capacity.

'Right, me old mate. I'm gonna need to get you into the back of my van with the minimum of aggravation and maximum consideration of appropriate noise levels for the time of night in a respectable residential area. So this time, I am gonna ask you politely to please drink this, and then a little later you'll wake up in a nice comfy cabin on the illy-alley-o. Alternatively, you can have more tough love from my business associate, Mr Spank, with the same result regarding your conveyance upon the briny deep, but a statistically larger probability of a fractured skull by the time you get there.'

Zal coughs and sighs, defeated.

'Tell Mr Spank to stand down. I'll play Alice.'

'Spoken like a gentleman.'

Greedy crouches down in front of Zal and holds the bottle in his left hand, unscrewing the lid with his right. He extends the bottle towards Zal's lips, at which point Zal grabs him around the back of the neck, pulling him off balance, and smashes his face into the iron pillar. Within a split second, Greedy is laid out flat with his hands cuffed together around the column in front of him. A second and a half after that, his ankles are cuffed together too, the chain looped through the iron frame of Zal's bed.

Zal pats him down, removes from various pockets the keys to the cuffs, van keys, wallet, phone.

Greedy splutters, spitting out some blood that has run from his nose into his mouth.

'How the fuck did you do that?' he asks.

Zal taps his nose.

'Trade secrets, *my son*.'

He flips through the guy's wallet, checks the information on offer inside. Albert Samuel Fleet, says his driver's licence. Zal takes note of a few things, then walks around into the guy's view. There's close to four hundred euros in cash, which Zal makes sure Fleet sees him pocket, before dropping the wallet contemptuously on to the floor. He doesn't particularly need it (though you can always use a little more); he just wants the asshole to know he's been burned.

Then Zal eyes the passport and tickets once more, and sighs.

He had told himself this prick wasn't going to make up his mind for him, but he already has; did so the moment he appeared. Zal's not going to Paris.

They murdered his dad, to get to him. They tortured his best friend, Karl, to get to him. Sooner or later, someone was always going to come after him, and to find him they'd go through whoever he loved. He couldn't face that again. He loved Angelique. There, that was his mind made up and admitted to. He loved Angelique. Loved her, how could there be any question? He'd spent only a couple of weeks with her, then the best part of a year thinking about her, defining his time according to her, trying to see a way of fitting her into the future picture. Loved her like he'd never loved any woman. Christ, you only had to look at what he'd done to be near her to understand that. He'd put himself in binary orbit around the policewoman investigating his own case, whom he'd taken hostage during the robbery, a material witness as well as the cop on his tail. All those added risks, just to spend time with her, just to find out who she was, just to find out if she felt the same way or if she was just playing him like she feared he was just playing her.

Yes he loved her, but nothing had changed: it was still impossible. The risk/benefit equation hadn't altered. He couldn't see how either of them could benefit, how they could possibly make it work, and that meant it wasn't worth the risk, especially when the risk was to her safety, not his. His coming back into her life could only mean trouble for her, danger for her. What happened to his dad, he had never entirely recovered from. If someone got to Angelique because of him . . .

No. For both their sakes, his mind was made up. He loved her, and that was why he was going to save her from himself. He wasn't going to Paris. He was going to disappear.

He picks up his passport, takes a bag from the wardrobe, starts throwing in what he'll need. He's leaving this place and not coming back.

'Wasting your time, mate,' Fleet says. 'Found you before, and I'll find you again. 'Less you're gonna kill me, that is.'

'No need, Albert, old chap. I know you'll come after me again. But ask yourself: what the hell makes you think it'll work out any different? Ask Hannigan what his track record is like against me. Look where he is now, he's in fuckin' jail because of me. Look where *you* are now.'

'*Pro tem*, mate. *Pro tem*. I'll find you, old son, you mark my words. Me and Mr Spank.'

Zal crouches in front of Fleet and smiles.

'Oh, I'm already looking forward to it. But I'll warn you just once: next time, I won't play nice.'

AUDITION

'You join me here on the red carpet outside London's famous—and utterly fabulous—Tivoli nightclub, for what I can promise you is going to be *the* music-biz bash of the year. Everybody who is anybody in the world of pop is going to be walking through those double doors in the next fifteen minutes, and you can also expect lots of famous names and faces from the world of fashion, from the world of football—accompanied, of course, by the very glamorous WAGs—from all the soaps, from comedy, you name it. It is going to be celebrity central here tonight, and the best part is that PV1 is your Access All Areas VIP pass: not just inside, but onstage, backstage and pretty much anywhere we can fit our camera, because we don't want you to miss a thing. My name is Jessica Hanson and I am *very* excited to be reporting for PopVision 1 tonight, *live*, from what will be an *amazing* party celebrating Nick Foster's Lifetime Achievement Award. As you know, Nick picked up the award itself at the Brits 08 ceremony just a few weeks ago, but tonight is the night when the pop world gathers round to celebrate what he has brought to the industry throughout a quarter of a century, during which . . .'

The girl babbles to camera with a strained giddiness, over-selling her enthusiasm and thus involuntarily conveying that she is precisely as excited as she is being paid to be. Girl? She's pushing forty, make-up not quite concealing the fan of lines flanking each eye, such marks of age seeming all the more pronounced by her trying to act like she's still twenty-two. She's buttonholing the new generation of spoilt nobodies while they take their time on the red carpet, as instructed by their PR people to ensure the paparazzi don't miss a shot. She's dressed in less material than you'd expect to find wrapped around cutlery in a decent restaurant, trying to look like she belongs among the lads-mag spank-bank nymphettes, but managing instead to resemble their embarrassing aunty, the one who is single, increasingly desperate and whose lack of a significant other is most manifest in her having nobody to tell her she shouldn't dress like that any more.

She's trying to sound pally with them but it's obvious they're only talking to her because she's got a cameraman behind her and a microphone bearing the PV1 logo. Poor cow, the folk she's interviewing were in nappies back when anyone gave a fuck who she was, titillating the post-pub audience late Friday nights on Channel Four. Her fifteen minutes long since expired and she knows it, but you gotta make a living, and they won't let you be merely a jobbing reporter on a shitty down-the-listings cable channel. No, you've got to humiliate yourself by pretending to still be a celebrity while you hold that mike, because in this world, there's no such thing as a mere reporter. As long as you're in front of the camera, you're a celebrity too.

The limos pull up, the snappers flash away. This is an indispensible assistance to Jessica and any members of the viewing public who have missed two consecutive issues of *Heat* magazine and therefore may not be sure which of these nondescript wank-stains are supposed to be famous this week.

A toothy adolescent, dressed in an outfit comprising what appears to be two curtain tie-backs, stops outside the vehicle she just exited and self-consciously adjusts the top half of her assemblage. This is ostensibly an effort to protect her modesty, but its true intention is to draw the paparazzi's communal attention to how little she's wearing. She's been sitting in a car for ten minutes, waiting for her turn to make an entrance, but it only occurs to her to check her tits aren't hanging out once she hits the red carpet? Sure.

Similarly, two teen members of a boy band whose new single has just tanked, spontaneously erupt into gesticulative argument in front of the cameras, the remaining constituents of their quartet pulling them apart and urging them towards the double doors. The single is called 'Not Pointing Any Fingers'. I said gesticulative, didn't I? Further description would be superfluous. It's fucking pathetic.

And in the gaps between the limos disgorging their payloads of ostentatiously attired affirmation addicts, Jessica fills the void with the media equivalent of a 69: interviewing another reporter.

'I'm joined for a chat just now by Damien Salter-Astwich, Fleet Street's Mister Showbiz. Damien, what can you say about Nick Foster? And does this crowd and this party not just say it all for you?'

Damien is some chinless cock-socket with unacceptably floppy peroxide hair and a retro-Sixties Italian boy suit. He's trying to pretend they were allowed a transistor radio in his dorm at Harrow.

'What can you say indeed, Jessica? The landscape of British pop music simply would not be the same without him. You're looking at three decades of, not just of hits, but of act after act after act, so many careers, so much success, and all one man. I mean, you go back to the Eighties and the guy was just a one-man hit factory, wasn't he? In those days, when you heard one of his songs, you knew within seconds that it *was* one of his, and that it was going to be a smash.'

'He really had the Midas Touch, didn't he?'

'He had something even more valuable, in my opinion: he had a recipe for success. As I was saying, just a few seconds into each song, it was instantly recognisable. Think of all those hits: they each started with a quick burst of synth drums, then straight into the chorus melody played right up front in the mix on synth brass, so that you already knew the hook by the time the first chorus came around on the vocals. Hit after hit, a winning formula, regardless who the singers were: it could be an established act or a complete newcomer, it didn't matter. Same style, same result: yet another smash!'

Their conversation runs over a compilation of clips showcasing the great man's multifarious musical progeny. The montage is so on-the-money as to strongly suggest this is a less-than-spontaneous little chin-wag. Jessica confirms this by teeing Damien up for his next verbal cum-shot.

'And what about all those spoilsports who said it wasn't real music? Because he took a bit of flak from the grumpy and serious types, didn't he?'

'Yeah, it's true some critics at the time said the songs all sounded the same and the performers were interchangeable, but if you ask me, Nick Foster democratised pop music, a lot more than we keep being told punk was supposed to have done. He demonstrated just how arbitrary notions of talent could be, proved that you didn't need angst, you didn't need posturing, you didn't need to have undergone an apprenticeship of rehearsing in filthy garages and playing in dingy little venues. You just needed a catchy tune and a bit of cheerful enthusiasm and anybody could be a pop star. I think he proved to everyone what he always said— that it's not about the music. It's about colour, it's about flamboyance, it's about fun. And that's why so many people are all here tonight.'

'Yeah, we're all here to have a *lot* of fun. There's quite a few of those Eighties stars have already walked along this red carpet tonight, but Nick was no slouch in the Nineties either, was he?'

'No slouch at all, Jessica. He made soap stars into pop stars, he made pop stars into TV stars, and he took the idea of a recipe for success a stage further in the Nineties when he started auditioning to put together the right personnel. In the Eighties, it was all about the formula for the songs, but in the Nineties, the formula he perfected was for the chemistry of the groups. That colour, that flamboyance I was talking about, he knew how to deliver that with a really vivid sense of variety by combining different images and personalities in a band. The result was that we got groups who offered something for everybody. Four Play had two boys and two girls, two blondes, two brunettes, two black, two white. Their biggest hit was "You're Dynamite", and they certainly were a pop explosion. Sunshine Seven must have been the most multi-ethnic line-up ever to grace the charts, and of course there were the Angel Cakes, the biggest girl group of that entire decade, maybe of all time. Even all these years after the Angel Cakes split, everybody can remember their nicknames— that was as important a hook as any melody, and Nick Foster was the man who understood that. He gave us five great characters to relate to before you even began to listen to a record. You had the funny one, Joker Angel, the wild one—'

Damien cuts himself off in response to the camera suddenly swinging away from him towards the kerb, where another limo has vomited out its payload of limelight-attracted moths: this time the recently re-formed members of Foster's late-Eighties abomination SWALK. The quartet split up in 1993 'to pursue solo projects'. These solo projects having largely transpired to involve point-of-delivery positions in the retail and catering industries, the group very quickly (but just a little too late) learned a new respect for each other's contributions, as well as an even greater regard for the creative input of the man who brought them together in the first place. Consequently, they are now attempting to ride the nostalgia wave surrounding all things Foster-associated, and hoping the clippings files don't go back far enough to exhume any of those nasty things they said about him taking all the credit and stifling their individuality.

Jessica manages to get one of them to join her for an interview. It's Baz, 'the cheeky one'. He looks less than delighted to have been buttonholed for a soundbite on cable TV while his bandmates are busy hawking themselves to the tabloid snappers. He rallies a bit, however, when Jessica's questions prompt the realisation that it's a chance to suck up to his erstwhile svengali, and cement SWALK's long-term association with the genius in the TV audience's minds.

'That's why we're all here, it's why we found so many people wanted us to re-form: the man has just given everybody great fun, good memories. A soundtrack to the happy times in our lives. I've had people come up to me and say they only need to hear "That Precious Look", which was our first number one, and they remember everything they enjoyed about that time. It came out on a vinyl seven-inch, but in a way the song itself became like a CD-ROM or a USB stick: it contains more than just a piece of music for people, it contains all these memories too.'

'Yeah, and of course, interesting you mentioned the USB stick, because Nick Foster has embraced the delivery technology of the Noughties in the most ingenious—not to mention controversial—way, hasn't he, with *Bedroom Popstars*?'

And now we really do get That Precious Look. It's a timeless moment as Baz stares wordlessly at his interviewer, licking and biting his lip like he's just eaten a doughnut.

'Yeah,' he eventually says. 'Amazing man, he really is. Look, I gotta catch up with the band— we've got a party to go to.'

With which he gets out of there fast before he says something he'll regret. Because even the no-talent, vapid, self-deluding forgettables who fronted Nick Foster's McSingles in the Eighties and Nineties struggle to conceal their righteous disgust, envy, resentment and gnawing sense of injustice whenever Nick Foster's pop-biz zenith is mentioned. It's like witnessing a guy who won a million on the football pools twenty years ago complaining about somebody trousering ten times that from the National Lottery. There's something going on there that's a lot more complicated than mere jealousy: a potent mix of insecurities, unpalatable truths and a dizzyingly vertiginous sense of fickle caprice. Small wonder Baz would rather not attempt to articulate it.

Damien, however, is seldom troubled by the problem of reticence. He talks shite for a living, and to him this particular subject is therefore like a society-wedding contract to a caterer.

'Well, Baz keen to reunite with his bandmates, something of a habit for him recently, you could say,' Jessica observes with, ooh, was that just the faintest trace of cynicism? 'He was out of time to talk about *Bedroom Popstars*,' she continues, 'but what's your take on it, Damien? Nick's finest hour?'

'Nick's finest hour, and finest half-hour-results update after the news,' he quips, in a way that makes you want to push the lenses of his glasses out of their frames and through both his eyeballs until the jelly squirts between your fingers and the jagged edges of the shards scrape the membranes lining the sockets. 'But seriously, *Bedroom Popstars*, what can you say? It was beyond genius. I mean, just when everybody thought the phone-in-vote talent format was starting to look a little worn, Nick found a way of pushing the envelope like it had never been pushed before. I was talking here just now about democratising pop music, well this was the ultimate democratisation of it, the ultimate

freeing up of the process by which anyone—
anyone—can become a pop star, if they want it
enough.'

'That is the key, isn't it?' Jessica interjects. 'How
much do you want it? Actually, we've got a little
clip here of Nick talking to us recently about the
idea and where it all came from.'

Now we're finally in the presence of greatness.
The man himself appears, reclining on a sofa
against a wall bedecked with gold and platinum
discs, his feet up on a table upon which several
award statuettes just happen to be nonchalantly
sitting among a number of magazines featuring his
face on the front cover. He's dressed in black
leather trousers, snakeskin cowboy boots and a
black t-shirt. Always a black t-shirt. There's no
colour more forgiving and concealing of middle-
age spread. He would be the living embodiment of
smug even if there was a bag over his head to hide
his face, but unfortunately we aren't granted such a
mercy, and his ridiculous but perpetually self-
satisfied visage smirks to camera in all its
improbably shiny glory. It looks like somebody
drove a stake through his scalp at the back of the
head and then twisted it until his skin was drawn
taut against his skull, which together with the long-
term effects of prolonged sunlamp exposure,
makes him look like a haemorrhoid with a
ponytail.

'There were a group of us in the VIP box at
Wembley, watching Robbie Williams, and I
happened to remark that his secret was that he
acted onstage like a teenager in his bedroom
playing at being a pop star, miming to the mirror
with a hairbrush for a microphone. That was the

joy he brought to his performance, and that was what people related to: he's not some distant one-in-a-million figure, a Freddie Mercury or a David Bowie, touched by genius and cultivated to the nth degree. He could be one of *them*, plucked from the crowd. And boom, that was the epiphany: when it comes to becoming a pop star, half the battle is simply *acting* like one. In that moment, you could say, *Bedroom Popstars* was born.'

An epiphany. Like in the Bible, he means. The annunciation. New Testament: Luke, Chapter One.

Bedroom Popstars, however, would better belong in Exodus, among the plagues of Egypt, or better still Revelation, as there can be fewer more gloweringly ominous harbingers of our world circling the drain than this latter-day Bedlam.

'Just so fresh, so audacious, so controversial,' Damien spunks. 'We'd seen *Pop Idol* and *The X Factor*, and these shows were simply about putting a voice and a face on a song: you didn't have to be able to play anything, you didn't have to write anything. Nick Foster took it one brilliant step further with a show in which you didn't have to sing either. As he said, it was about *acting* like a pop star: who was the best at miming, posing—vogue-ing, if you like—to these catchy Nick Foster songs. And it had such a symbiosis with the technology of today, in that when the singles came out, they were on USB sticks, because it was an inextricably audio-visual entity. You weren't just buying the song, you were buying the winning act's performance, which included the video, their early attempts on the show itself, plus all sorts of personal extras.'

'He's always been one step ahead of the game, hasn't he?'

'And one step ahead of the critics, Jessica. They were outraged, they all said that what he had done was to prove the music was incidental—incidental music ha ha ha—but in fact what he had proved was that it was about *more* than the music. He started a real debate about this, about what it takes to be a pop star. It used to be the argument that the so-called manufactured acts weren't as talented as the muso-songwriter types; then the argument became that at least the manufactured singers were more talented than these new acts just *pretending* to be pop stars. But it comes back to Nick Foster watching Robbie Williams and realising there's a talent to acting like a pop star, there's a talent to entertaining people through putting on the kind of performance that is expected of being a pop star. But of course, the most controversial aspect of the show—and for my money the most fun—was that it required that talent, that performance, to transcend the stage and the spotlights and break into day-to-day life.'

We cut back to the haemorrhoid again.

'In my three decades in this business, one thing I've learned is that people who want to be stars don't care how they get there. The ones I've got the most out of were the people who didn't particularly care whether it was music that got them there: they'd have as easily gone with acting, TV presenting, just whatever opportunity knocked. They know that it's fame they want, and they will do whatever it takes to get it.'

Which is our preface for a new montage, a veritable kaleidoscope of tabloid front pages and

113

glossy magazine splashes harvested from the antics of the *Bedroom Popstars* contestants throughout its two appallingly successful series. Nightclub fights, drugs, booze, outrageous statements, phone porn, affairs, orgies, all manner of sexcapades.

'The programme itself was actually the least part of the package, almost mirroring the way the song itself was the least part of the USB single,' Damien enthuses. '*Bedroom Popstars* became an even more twenty-four/seven phenomenon than *Big Brother*. The contestants knew that they weren't just competing on stage on a Saturday night—they were competing to see who could do the most to keep themselves in the spotlight through the week, between shows. Who could get on the front pages, the news bulletins, the websites. Who could act like a pop star offstage, even if they were a hard-up sales assistant or nurse or delivery driver. I mean, this is why Nick Foster is more than merely a pop svengali figure. He has created a non-stop form of public performance, people turning themselves into a continuous, *living* pop exhibit for our ongoing entertainment.'

Jessica thanks Damien for his contribution. Nick Foster ought to be thanking him too: after such a sustained and enthusiastic rimming session, his sphincter must be looking almost as gleamingly polished as his implausibly shiny head. However, Nick has other more pressing things to be concerning himself with right now.

'Okay, get your dancing shoes on at home,' Jessica gushes, 'because it's our turn to walk up the red carpet as PV1 takes you, right now, into a packed Tivoli, where the party is kicking off with a *live* performance from none other than this

114

year's *Bedroom Popstars* winners: yes, it's the incomparable, the unique Vogue 2.2!'

The camera swoops past Jessica and zooms between two dozen packed and heaving circular tables, up over a pulsatile miasma of sheer fabric and tit-tape on the dancefloor, to focus finally upon a dry-ice shrouded dais flanked by two stacks of speakers. Prowling this befogged promontory are the three proven, international-class media whores comprising Vogue 2.2: Sally, Anika and Wilson. They strut back and forth in energetic but yawningly hackneyed choreography, pumping thick radio mikes in each of their right hands like weights at the gym, an inexplicable accessory as it is a matter of pride, never mind mere fact, that they won't be making any vocal contribution requiring electronic amplification. Their job is just to stomp around in synchronicity, throw some shapes, exhibit limited quantities of an indefinable property referred to as 'attitude' and, of course, mime (though truth be told they aren't even required to do much of that, the accompanying 'song' largely comprising sampled loops of previous Nick Foster hits).

If anyone had any doubt over Foster's contempt for the contestants (or the idiots watching), then it should have been erased simply by the name he gave his latest creation: Vogue 2.2. That's as opposed to last year's winning line-up, who were called Vogue 1.1.

Vogue 1.1 were a five-piece . . . (I'm unsure of the appropriate collective noun, as outfit, line-up, group and band all afford too much dignity-by-association through having been applied to people who actually made some kind of net contribution

115

to the human race. Let's go with ring. As in piece. As in gaping. As in Goatse.) Vogue 1.1 were a five-piece ring, but Foster wasn't so unimaginative as to entirely repeat the formula. *Bedroom Popstars* series two was, he initially said, going to create another five-piece ring, and the show spent a couple of months whittling the contenders down to that number, before he pulled a ratings-rocketing flanker and announced that the series would be extending by a month, and that the magic number had changed to three. However, instead of also extending the usual voting and elimination process, Foster informed them that this final streamlining would be conducted on a voluntary basis.

Two people would simply have to decide fame and pop-stardom weren't for them, and resign from the competition. This ostensibly democratic invitation to reflection and, ultimately, selflessness, was in fact the equivalent of dropping the five of them on to a rubber dinghy with five knives, three lifejackets and one fucking big puncture. It was also a coded signal to the tabloids that it was open season.

Foster defended the ensuing four-week cross-media backstabbing, secret-splashing, character-shredding bloodbath by saying that it was nothing they wouldn't be expected to endure as pop stars and thus it was a crucial part of the audition process: 'It's one thing proving you've got what it takes to attract media attention—this will prove who's got what it takes to really *handle* that attention.'

Consequently the three survivors had demonstrated themselves to be in possession of formidably thick skins, in combination with merciless predatory instincts and an almost total absence of self-awareness. This is just as well, given the volume of latent resentment that lurks behind so many of the smiles being displayed for the cameras tonight as Vogue 2.2 greedily guzzle the limelight. The stars around the tables are no doubt telling themselves their barely suppressed hatred stems from their own true talents being insulted by this sham, but secretly every last one of them must have been cut deep by the awareness of what Vogue 1.1, 2.2 and the whole *Bedroom Popstars* phenomenon implies about their own successes.

Vogue 2.2's number comes to an end and they strike a frozen pose as they soak up the cheers from the blanks (i.e. non-celebs) on the dancefloor, and the conspicuously strained applause from the tables. The latter constituency clap through teeth-gritted smiles, which underlines their impotent compliance: if they truly wanted to make a point, someone should have stood up and asked the cunts for an encore: then they'd really be fucked. However, nobody wants to shit where they eat. They are all here to suck up to Nick Foster in the hope that his blessings will continue and thus sustain the magic spell that has granted them places at those round tables, rather than hoping to get noticed down on the dancefloor, or outside in the street with the utter nobodies.

What a gathering, truly. I can't look at the dry ice without thinking of just how many of these parasitic nonentities I could have taken out in one go. I did that once to a club full of neo-Nazis watching a group called Krystalnacht: got fifty-five of them, including the band, by substituting the dry ice with poison gas. Very disappointingly, nobody in the media seemed to get the joke. You really have to spell things out to those bastards, because if you don't, someone else will, and it's their spin that will colour the story. That's why, tempting as it was, I knew that a similar stunt here would only backfire. The last thing I wanted was to make every one of these spoiled puke-bags immortal, and have *all* their fucking dismal records re-released.

Besides, I'm not a monster any more. Well, yeah, I probably still am, but I'm a monster with some panache. There's no style or flair to such indiscriminate killing, and no moral to the tale either.

Vogue 2.2 exit the dais, replaced by the evening's mistress of ceremonies, Mica Dahl, erstwhile 'Cheeky Angel', subsequently lost to the music world, though her departure has been television's gain these past couple of years as the main presenter of *Bedroom Popstars*. A catwalk suddenly extends phallically from the dais on to the dancefloor and she slinks along it in a pair of knee-high boots more normally associated with the principal boy in a panto. A podium erupts from the floor at the end of the catwalk as she approaches, while ten yards behind her, a huge blue panel reveals itself to be a projection screen as a still image of the haemorrhoid's taut hide becomes manifest upon it.

She says a few words, bids everyone welcome. Something, however, isn't quite to spec. You'd have to be looking for it, but it's there. She's got the practised professional grin switched on when she's talking, but as soon as she's introduced the first montage on the screen and she thinks the camera is off her, you can see the uncertainty on her face as she presses the earpiece and listens for instruction.

The montage ends and she turns the smile back on. She cracks some jokes, makes specific mention of a few of the assembled stars, then introduces the next live performance, this time from re-formed Eighties Foster-pop veterans No False Moves! (the exclamation mark being an incorporated part of their handle). It's an atypically slow number that they give us; a hit in its time, but not the high-energy gay-disco fare they are remembered for. Similarly they have opted to dress in identical long coats rather than their contemporary preference of skin-tight day-glo spandex. The choice of track shares its explanation with the choice of wardrobe: it's all to disguise that the act people remember as four hyperactive androgynous teenagers are now just four shuffling fat blokes.

Four Gentle Moves having been spent, we're back with Anxious Angel, who announces a quick break, cueing spotlights on the dancefloor and music on the speakers.

There is, if you look beyond the gloss and gels, a bit of a ferment about the place: much running around beyond the sidelines, inconspicuous shuttlings by personnel in official t-shirts and headsets.

Jessica and co have got wind of it, and bearing in mind the threat of her attention-deficit audience switching over if she doesn't keep hitting their id-gratification feed-bar, she directs the camera away from the dancefloor and back to herself.

119

'Bit of gossip for you folks, just between you and me, the rumour around Tivoli right now is that the guest of honour might be fashionably late for his own party. You've seen Mica up there, well I can tell you that she was supposed to be introducing Nick at the end of that first montage, but as yet there's no sign of him. You're actually getting a rare insight into how these things can be. It's the swan effect, you know: all smooth and graceful on the surface but everything frantic beneath. If you ask me, though, this is just more exquisite showmanship from Nick Foster: he's keeping everybody guessing, making sure nobody takes him for granted. Dancing to no-one else's tune, working to his own sense of timing, and when he finally makes his entrance, you can guarantee it'll be a show-stopper.'

Quite, Jessica, quite. A show-stopper, undoubtedly.

There's a guy in a booth inside the Tivoli, sitting at a console watching monitors and computer screens, looking at different camera views, video cues, track lists, a scrolling script. His name is Garth Whiting. He is directing the show. Probably getting more than a little jumpy at this point, what with his presenter unaccustomedly winging it and the evening's star attraction as yet unaccounted for. Garth hasn't been able to reach him all day: mobile, landline, email, texts, nothing. Fucking prima donnas, huh? Not to worry, though. I'll soon be sealing off the main source of his anxiety, if not exactly putting his mind at ease.

I'm sitting looking at just the one monitor myself, less than a hundred yards away, as it happens. I've got a digital receiver—'TV on a

120

stick!'—plugged into a USB port on the laptop, and PopVision 1 open in one window. Like Garth, I have alternative views available to switch to, one of which will soon be available to him also.

I send him an email, then call him on his mobile to alert him to this. A couple of minutes later, technically Garth is still controlling the show, but it would be fair to say he's got a new producer. I've changed the aesthetic but I'm a hands-off guy to work with, restricting my input to two brief instructions.

One: switch the screen above the dais to stream from the link in the email.

Two: the second you cut it off, he dies.

There's a few seconds of silence after the musical interlude finishes, with no sign of Mica returning to the end of the catwalk. The lights go down, however, and the screen above the stage has gone enticingly blank. There's a strangely unprompted silence as a sense of anticipation runs through the place.

'I think something special might be coming up,' Jessica guesses with a wink.

You go, girl.

Suddenly, a bass drum begins to pound, and a silhouette of a male figure in classic Bond pose appears on the big screen. There's a cheer as the silhouette becomes full colour, revealing an old *Radio Times* cover image of Nick Foster holding a TV remote like it's a Walther PPK. This heralds a sequence of familiar publicity stills of the guest of honour, changing every few beats.

A guitar starts to play above the bass beat, a dirty death-metal-style snick-snick creeping along ominously like an anaconda through the undergrowth.

Captions begin to appear, white out of black, in between the changing photographs of Foster.

'A man of many faces.'

'Many styles.'

'You've seen him a thousand times.'

'You've heard him a thousand times.'

Bass guitar augments the rhythm section, broadening the sound, building up to something. In the audience, people begin to clap along to the drumbeat.

'But you've never seen him . . .'

'You've never heard him . . .'

A keyboard chimes over the guitars: synth brass, classic Nick Foster Eighties style. It's playing the chorus of the chart-topper he wrote for one-hit wonder Wendy Clear, 'Hurts So Good', but it's a twisted, discordant version of the melody: it's recognisable, just, an even more nightmarish migraine-warped rendition than anything Marilyn Manson did to 'Tainted Love' or 'Sweet Dreams'. Sounds pretty good to me, but then I'm biased. I recorded it.

'Like this . . .' concludes the caption.

And with it, finally, a vocal sounds over the screen, now faded to black. It's a scream: a throat-rasping, lung-deflating hardcore-metal graaawl that Jonathan from Korn would have been proud of. It sounds like the vocalist is having a telegraph pole slowly removed from his arse, footpegs intact, or perhaps more accurately like he's been wired to the mains. Very, very accurately, in fact.

On the screen, the all-black view tracks out from what is revealed to be a pupil, revealed slowly to be the centre of a bloodshot, scared and terrified eye, then further revealed to belong to one side of an implausibly taut and shiny face.

A second vocal plays over the music: my own, telephone-distortion take on the Eighties hit chorus:

> Hurts so good. Hurts so good.
> Turn up the music cause it
> Hurts so good
> I need your lovin cause it
> Hurts so good
> I know we shouldn't but it
> Hurts so good
> Please don't stop, you know it
> Hurts so hurts so gooooood

At the end of the chorus comes another scream, strangely in key, though this is because I sampled it and made a few adjustments. And as it plays, the image tracks back further to show Nick Foster lashed naked, Prometheus-style, to a gigantic Marshall amp, into which is plugged an equally gigantic black guitar lead.

The amp is not real, just a scaled-up façade, the main part constructed of a highly conductive metal mesh, with the logo and dials printed on thick card. Face-on, as seen upon the screen above the stage, it looks surprisingly convincing, with the weird scale effect that the amp ceases to look giant and instead Foster begins to look miniaturised.

The guitar lead is not real either, just a length of black rubber tubing.

The electricity *is* real. So are the screams, though as I admitted, they have been manipulated to sound in key.

123

The tune plays on through another round of the chorus. Most of the audience keep clapping. They think it's a gimmick. Even Jessica and PV1 stay with it, though their director has the presence of mind to digitally scramble the pixels around Foster's nads. (The genitals stay censored throughout the three minutes of torture they unflinchingly relay to the nation before finally pulling the plug on the transmission when Foster pisses himself. British values, gotta love 'em.)

And with the guest of honour finally having been presented to the crowd, it's time for them to meet their real host. An inset window on the screen shows my hand poised over a voltage control and a charge button.

'Ladies and gentlemen,' my voice hails out crisply from the PA. 'We are here tonight to pay tribute to the contribution to British music made by the one and only Nick Foster. The stars are here and I'm sure all of them have their own memories and impressions of him . . . but what does the man himself think of his own legacy?'

A second inset appears on the screen, showing hand-held footage of the same room in Foster's house as he had posed in for the clips shown earlier on PV1. It pans across from the framed gold discs to a wall housing a good few hundred CDs.

'This is Nick's own CD collection,' I explain. 'This is a unique insight into what fuels and inspires a pop genius. And it's a mark of his well-known modesty that these shelves play host to absolutely *none* of his own recordings. His home PC and his iPod tell a similar story. Not one Nick Foster track appears on his personal playlists. But stranger still, there doesn't appear to be any manufactured disposable pop records by anybody else in here either. Why would he deny himself the very pleasure that he brings to millions?'

The screen flashes up a caption at this point, white out of black:

'My old dad worked down Billingsgate fish market, and that's how he learned the difference between a tradable commodity and something genuinely valuable. There's fish that's just for buying and selling. You don't fucking eat it.'
—Nick Foster, *Melody Maker*, August 1992

'On a night like tonight, it would be great if we could appraise Nick's legacy by exploring just how ingrained his music has become in the national consciousness. Admittedly that's a difficult thing to measure, but just for fun, we're going to see how deeply Nick's music is ingrained in his *own* consciousness, and you at home can see how your answers compare against the man himself. I should add that if you really want to join in properly and get the full effect, you're going to need two crocodile clips, some electric cabling, and a variable resistor to regulate the current.'

The screen now shows a close-up of Foster's sweating and contorted face, my hand and the electrical controls still visible in the inset window. My hand twists the dial until the needle reads 1,000 volts.

'What's going to happen here is that I'm going to ask Nick to play name that tune, but just from the lyrics. If he gets it right, I knock five hundred volts off the next shock. If he gets it wrong, I let him hear the melody, which gives him the chance to win two hundred volts off. It's not as much, clearly, but when it's pumping through your unearthed flesh at five amperes of current, you'll be grateful for it, believe me. Oh, yes, and before the shock, there's a bonus question, but we'll get to that in time.

'Okay, first lyric for you, Nick, and for everyone watching at home. Here we go: "Gonna make my move, gonna feel the groove. Gonna feel the heat, gonna dance the beat." For five hundred volts, name that tune.'

He says nothing. The camera tracks out again, showing all of his naked body lashed to the giant amp. I repeat the lyric.

'I'm going to have to hurry you, Nick. I realise the lyrics are utterly vacuous and probably indistinguishable from a thousand other shitty songs you wrote, but *you did* write them, so why not take a stab. I'll even give you a clue: it was from 1987 and it reached number three in the UK charts.'

'It was . . .' he croaks, then has to swallow, his mouth too dry to speak. 'It was "Here I Go Tonight".'

'Ooooh, sorry, Nick, that's not the answer I've got here. But there's still two hundred volts at stake, so why not have a listen to the melody and have another go at naming your own number-three hit single from 1987.'

The trademark synth brass intro plays over the PA. Nick's face strains, difficult as it might be to imagine his facial muscles stretching any tighter. The music fades out before the vocal can begin,

126

though it's my firm belief that it wouldn't help. You can see his head jiggle as he tries to keep the song playing in his head.

'Is it "No Stopping Me"?'

'Is it "No Stopping Me"? I wonder what everybody is saying at home.'

Around the Tivoli, where they still think it's some kind of risqué joke, there are a lot of shaking heads. PV1, still broadcasting at this point, have even located and zoomed in on Sandi Bay, the woman who sang the song I'm looking for. She bopped around a swimming pool in hot-pants and a boob-tube in the video, but right now she looks like a deputy bank manager on her works night out. She's also looking anxious and confused, wincing uncertainly at his wrong answer.

'I'm sorry, Nick. The answer is "Dance All Summer Long".'

He closes his eyes, grits his teeth, bracing himself for the electricity.

'Now, now, relax. You didn't get any volts off, but there's still the bonus question. It's worth two hundred and fifty volts, but the catch is, if you get it wrong, those two hundred and fifty volts get added, not taken away. Now, I've already given you the name of the track, and the year and the chart position, but for two hundred and fifty volts, can you tell me who had a hit with it for you?'

He's breathing quickly, over-excited, panicky. Sounds like he has the answer. On PV1, the camera remains fixed upon Sandi Bank Manager.

'C-can I pass?' he asks.

'No, Nick, I need an answer. Have a think: 1987, top three, "Dance All Summer Long". Summer, think summer. There's a clue.'

He nods, blinking, bites his lip.

'Okay, okay.' Swallows again. 'It was Surfs Up!'

'That would be Surfs Up!, spelled with no apostrophe before the "s" and a nauseatingly jaunty but redundant exclamation mark after the "p", yes?'

'Yes,' he affirms expectantly.

Sandi Bay looks a little grey, her cheeks sucked in. She's oblivious of the camera, but she doesn't need to know anybody's looking in order to feel humiliated. There, there, pet, it was a long time ago.

' "Dance All Summer Long" by Surfs Up!. Let's just hear a little more of the song and find out if you were right.'

The track fades up again, in time for the vocal to start. 'Gonna make my move, gonna feel the groove.' It's an unmistakably female voice. While none of Surfs Up!'s three personnel were ever likely to be confused with Mark Lanegan, they were nevertheless a boy band. From his face, it's obvious that Foster remembers that much at least.

I hit the button and bring it home to the Tivoli crowd that, while I'm finding it funny in the extreme, it's no fucking joke.

Before PV1 drops the feed, they pick up one final shot of Sandi Bay. She's not looking quite as horrified as everyone else might deem entirely appropriate.

The Tivoli itself, with Garth at the helm, doesn't have the option to cut the feed. With no PV1, I don't see what happens inside, but I do see people begin to exit the place within a few minutes, and I know from the laptop that the feed streaming my recording of Nick Foster's gameshow experience keeps running until it loops; in fact, it is a good few minutes into repeating itself before anybody, despite deducing that it's not live, dares to risk pulling the plug. This means that the ensuing and increasingly hysterical abandonment of the evening's activities is carried out against the backdrop of Nick failing to identify his 1994 number-one smash 'Party Round the Pool' from the lyrics: 'Power up the lights, Get the music playin' loud, Get your baby movin' right, Cause it's just a crazy crowd'. He subsequently nails it from the intro before blotting his copybook again by attributing it to Sunshine Seven instead of Four Play. Easy mistake to make when there is nothing to musically distinguish one set of willing fame-ravenous whores from another, but it still cancels the two hundred and fifty volts he thought he had bought back.

Bzzzz.

Hurts so good. Hurts so good.

Turn up the music cause it

Hurts so good.

The ensuing performance outside is considerably less of a choreographed spectacle, but undeniably a spectacle all the same. Jesus God, have you ever seen an emergency celebrity-evac? All of a sudden there's a hundred self-important wankers trying to occupy the same stretch of concourse, determined not to spill beyond the awning-covered red-carpet area on to the ordinary, proletarian-accessible thoroughfare as they look expectantly for the limo they—or their 'people'— are desperately trying to hail on their mobiles. There's a unique expression comes across most of their faces, one that really deserves a coinage, ever after commemorative of this night: it's a combination of confusion, anxiety and affronted indignation as they each assess the ever-complicating situation and the realisation dawns on each of them that poowittuwme won't be getting any special consideration. The phrase 'Do you know who I am?' already forming in their minds, gets forced back unspoken and they simply have to take their chances the same as everybody else.

They're largely spared the paparazzi, at least for a few minutes, the snappers having mostly fucked off for a drink and a bite in the time between the grand entrances and the generally happier hunting to be had shooting the better-lubricated exits. A couple of keener ones are back in position but not yet set up again and consequently taken by surprise by the sudden swell of tux and taffeta.

The limos start to pull in, but there's a bottleneck created by there being only enough space for two to park at once, and the first one there, being mine, has further reduced the pavement space. However, with the departing celebs seemingly determined not to be seen venturing desperately along the street in search of their rides, lest they appear like common citizens attempting to hail a Hackney, the resulting melange beneath the awning soon resembles the world's most over-dressed late-night taxi queue. There's even young males squaring up to each other and girls shedding mascara-smearing tears. Throw in a few kebabs and the picture would be complete.

My limo gets a lot of hopeful glances, though the absence of a uniformed driver getting out to hold open a door for the VIPs soon prompts sufficient scrutiny for them to notice the printed sign occupying the rear passenger-side window. 'Vogue 2.2' it states. The limo is not a stretch, but a customised Merc with rear- and front-facing seats in the back. Next to so much vulgar ostentation, it subtly whispers both class and 'fuck right off'.

Most of them clock the sign and begin looking along the growing line. What's just precious is that even if they spot their driver back in the queue, they don't move towards him for a quicker getaway, especially not with the photographers increasingly geared up. One, however, does finally approach my car and knocks on the front passenger-side window. I ignore it. He tries again, eliciting the same lack of response. Outraged by this, he comes around the front so he can see me through the windscreen. I point to a second 'Vogue 2.2' sign on the dashboard.

'We're with the same record label, mate, and we're going to the same fucking place. We'll all fit, so open up, all right?'

I just shake my head. He doesn't like this. He slaps the bonnet angrily.

'Listen, mate. Don't give us any jobsworth shit or this'll be the last shift you work for this firm. Is this face not slightly fucking familiar? "You're Dynamite" ring any bells? I'm Daz Hartnell, we're Four Play, and this limo is going to the Tailspin Records reception, so if you don't want your P45 tomorrow morning, open the fucking door.'

I glance back into the passenger section. He's right. It's officially built for six, but there's plenty of room for everybody.

I open the fucking door.

131

The Vogue 2.2 trio pile in right at Four Play's backs, trying to disguise their indignation at this imposition. Fake camaraderie at their shared shock and discomfiture is exchanged as they take their seats, the Vogue girls tutting a little at having to squeeze between two Four Play counterparts already occupying the forward-facing row.

I pull away into the London night and everything calms down soon after. The sleep agent sees to that.

*　　*　　*

It's only music. That's what people always say. Don't take it so seriously. It's only music.

Anyone who says that deserves to listen to nothing but Nick Foster's aural chewing gum for the rest of their lives. Music is one of the achievements that truly elevates man above the beasts, one of the ways in which he gives expression to something within that is greater than himself, older than himself, more universal, more enduring. Music makes manifest the aspiration of the human soul. Or at least it should try to.

'Only music' is an oxymoron, like 'only life', 'only oxygen', 'only your firstborn'.

'Don't take it so seriously. Each to his own. It's all a matter of taste.'

Away and throw shite at yourself.

This is not about taste. That, in fact, is part of the fucking point. It's about a lack of respect for music itself. That's what all these little fame-whores are guilty of: Four Play, Sunshine Seven, Surfs Up!, Angel Cakes, One False Move, the whole Foster roster, every manufactured plastic turd ever shat out of the *X Factor* anus, every auditioned, formularised and split-second synchronised troupe of interchangeable grinning marionettes. If I simply didn't like their music, I wouldn't be so offended. Let me state again: this is not a matter of taste. What offends me is that these eager little slut-monkeys are not interested enough in music to particularly *have* a taste. They'll sing anything they're told to, they don't care what. They're not *interested* in music. They're only interested in locating the path of least resistance between themselves and the cover of *Heat* magazine.

Bedroom bastarding *Popstars*? The most laughable irony about this cultural atrocity is that when they were kids, the contestants on this partial-birth abortion of a show *never* danced around their bedrooms playing air guitar or singing into their hairbrushes. These little narcissi would instead have been posing for imaginary photo-shoots, calling their pretend publicist on their toy mobile, and practising climbing out of their pedal cars in front of imaginary paparazzi.

It's not about the music for them. When I cased Foster's CD collection, I wasn't remotely surprised that it didn't contain any of the vapid, disposable, prole-pod fodder he inflicts on the rest of us. But what surprised me even less was that, despite having millions to spend and CDs flung at him for free anyway, his collection was no bigger than you'd find in any middle-class, middle-aged suburban living room. I owned more records than that by the time I graduated uni. Music is merely his business, the field he works in. He has no love for it, never mind passion. There's fish that's just

133

for buying and selling, he said, but the truth is the cunt doesn't even *like* fish that much.

I always loved music. I always had a passion for it. Maybe too much of a passion; certainly far more passion than the people I tried to work with could ever handle. Yeah, that was the problem: I can appreciate that, now that I've got the perspective afforded by several years and a couple of lifetimes' distance. At the time, all I could see was their limitations, their pettiness, their mediocrity. Now I can see that I presented too much of a challenge. I brought too much passion, too much vision. They realised they were in over their heads, I wasn't another teenage dreamer simply *playing* at being in a band, and the difference between my own abilities, my own scope and theirs was what forced that realisation upon them.

At uni, the fools I fell in with really *were* just bedroom pop stars; calling them a garage band would be granting the sorry undertaking an unmerited sense of ambition. They only wanted to pretend, wanted to play covers, close their eyes and merely imagine themselves in the role of Stuart Adamson, Mike Scott, Billy Duffy, whoever was cool that term. To think that they ridiculed me for wanting to bring my own uniqueness, my own expression in from the start. The Bacchae: it could have been beautiful, should have been triumphant. I wanted to give them a name, a sound, a message, a uniqueness, that would have them striving for the heavens from the off. No acting like somebody else, no borrowed, off-the-peg costume of what a band should be, but a distinct, strident vision, constructed uncompromisingly, unflinchingly and unapologetically from the ground up.

They couldn't see what a gift I was laying before them. Said I was trying to fly before I could walk. I was ready to soar, motherfuckers. Wasn't my fault they were still insecure about every step they took in their babywalkers. Probably for the best that the band were strangled at birth, imploding after one disastrous gig, Larry the Little Drummer Boy being particularly culpable for the fiasco due to his insistence on taking lead vocals from behind his kit on one of the songs.

I found a more technically capable collective to work with a few years later, post-uni, and really thought that our mutual maturity as well as more comparable musical gifts would make for a more stable platform. Unfortunately I encountered merely different degrees of the same problem. They had the ability the student wannabes lacked, but were just as scared of my passion and intimidated by the scope of my vision. I laid my promise before them and they recoiled in fear, again unsettled by confronting the gap between what I thought a band could do and the limits of what they believed themselves capable. There was the petty egotism to deal with too, folk who just couldn't accept their station. They rejected my songs and my ideas because it would have meant playing second fiddle: they saw immediately that they were dealing with something unique, and which could not be diluted in the name of collaboration. They preferred the prospect of first-name writing credits on something inferior, to mere contributor status on what I would have coaxed out of them. Thus they rejected the chance of uniqueness and settled for being another copy of something that was already over-familiar.

Admittedly it was a better copy than the student clowns could manage, and, calling themselves Chambers of Torment, they did go on to sell a lot of records, but they never did anything unique, anything truly original. I always laugh when I see them described as 'fearsome' or 'scary', just because of a few tattoos and some hackneyed horror imagery in their videos. I know the truth. I know they are cowards.

But that's not a crime. I don't hold anything against them. I've mellowed. I know now that I was asking too much, asking them to become something they never had it in themselves to pull off. I know they are essentially just a facsimile of the groups they looked up to; really just a more sophisticated version of the covers band that my student colleagues were happy to settle for being. But they are playing what they always wanted to play, it's their taste, their chosen style—and they do bring to it a love, a passion. I can respect that.

Nick Foster, however, never brought any love or passion to the music he produced, and nor did the Ritalin-deficient limelight-junkies he employed. Under Foster's instruction, they never put any heart into it, never mind soul. He ensured there was never any real expression in their vocals, no unleashing of turbulent emotion. Too much expression, too much emotion in the vocal and you can frighten the horses. It can have an adverse effect on your chances of securing high rotation on the pop stations. Can't have something like that playing in the office of an afternoon, or you might disengage the drones from their allocated tasks, and, God help us, you might accidentally revive their anaesthetised imaginations.

Never any passion, any expression, any genuine, unfettered emotion on a Nick Foster record. And never any Nick Foster either: not so much as a backing track. Couldn't sing a note, apparently. (Perhaps that was why he was so dedicated in suppressing the expression of those who could.) But both of those things are about to change. I'm going to release—digitally, of course—my new version of 'Hurts So Good', as debuted at tonight's party. And on it, as you heard, is guest vocalist Nick Foster, giving vent to more expression than he's ever loosed in his life; unleashing more emotion than all the singers he's ever produced have cumulatively committed to tape.

I know we shouldn't but it
Hurts so good
Please don't stop, you know it
Hurts so hurts so gooooood.

*   *   *

Oh, wait, you must see this, it's one of my favourite bits.

You really ought to download the whole thing if you haven't already. I know they keep taking the links down, but this is the web, for fuck's sake, who are they kidding. The internet treats censorship as damage and routes around it. There's new links going up all the time. If you can't find them because the cops or the record companies have nobbled Google, then look for the news groups. The files are on alt.blackspirit.binaries and the chat is on alt.blackspirit.discussion. If any particular files aren't visible, just request a pea-roast. (You won't find the 'Hurts So Good' mp3 there, though, as it ain't free. I won't see any of the money, so don't worry your hypocritical little conscience about that, but it only counts towards the chart if it's a paid-for commercial download, so I had to rig something.)

I love the choreography here, the way they're all spaced out and in position in the classic Four Play video line-up: black boy, white girl, black girl, white boy. The giant Marshall amp in the background adds a kind of synergy too, like when the same iconography appears on all the videos and artwork on multiple singles from the same album. The synths are new; well, new for any video Four Play have ever lined up in. The headset mikes are just precious too—they strike precisely the right late-Nineties note.

I'm really proud of this: absurdly, gigglingly so. I'm not normally impressed by the supposed merits of improvisation. Never understood why anyone ever thought it was a worthwhile form of comedy. Just because you think up some shit gag on the spur of the moment doesn't transform that same remark into something funny. My sense of humour doesn't work that way—am I supposed to set my laugh sensitivity somehow lower, recalibrated to make allowances that I don't need to make for witticisms that somebody actually took some time and care to construct? But this, even if I say so myself, is something of a triumph of on-the-hoof thinking, proof that necessity really can be the mother of invention and all that.

I already had plans for Vogue 2.2, and reckoned I could incorporate the extra passengers into the same scheme, but as I drove my cargo of somnolent stars beneath the darkness, I was struck by inspiration. I can't claim I pulled it entirely out of nowhere, though: there was a traceable genus to this idea. Young Daz gave me it, having reminded me of Four Play's biggest number-one hit.

The headset mikes are non-removable: they're attached to steel collars around their necks and padlocked at the back. I attached one to a crash-test dummy before we started recording, to ensure they gave me a more committed vocal performance than Foster ever coaxed out of them. There's a small plastic-explosive charge attached to the crosspiece supporting the microphone's pick-up head, the device radio-controlled by me. The charge isn't going to demolish any buildings, but as I demonstrated to them, was quite enough to obliterate a dummy's head. Producers sometimes have to do extreme things to get the sound or the feel they're looking for.

They stand at the keyboards, in formation, singing their greatest hit and thus adding a new vocal track to my evolving Nick Foster tribute song. It wasn't much of a challenge to mix the Four Play number into 'Hurts So Good'—after all, you could run most of Foster's output together and never notice an edit. Same time signature, same production, same horrible sampled snare. With these latter two encumbrances removed, however, it's starting to sound pretty interesting.

Secrets of the successful pop producer and video director: they've been told that if they don't sing when they're supposed to, they die. If they move away from the keyboards, they die. But as with the Daddy, I've given them an out. No point in punishment without the chance of redemption.

There's a synth-brass break before the final

chorus, a structural staple of every Nick Foster hit. It just reprises the opening run-through of the melody before the vocals join in for the final chorus, usually repeated to fade. Outside of that fucking four-note Intel signature, it's probably by some distance the single piece of music these four have heard the most in their entire lives, so we're not talking Geoffrey Rush in the movie *Shine* here. When that part of the song comes around, if any of them can play it on their keyboards—not note-perfect, just recognisably and in key—their collar comes off. If they can actually play *one* little bit of *one* song, the song that made them stars, then I'll sedate them again, drive them away from this place and leave them somewhere they'll be found, unharmed. Only Daz has seen my face, and I assure him he hasn't seen what he thinks he's seen.

There is, as promised, real feeling to the vocal. The girls are actually crying as they sing. Sweet emotion, as Aerosmith put it. They're still not very good, but at least they're putting their hearts into it at long last, and that's all you can ask. I always knew they had it in them.

We go verse, chorus, verse, chorus, then into that final verse before the crucial synth break. Can they play it? That one little melody? To save their lives? Can they?

The ensuing synth-mashing cacophony says no. That part will not make the final audio mix, though it stays in the video for obvious reasons.

They're all in tears now, but what troopers, what pros: they do as they're told and keep singing on into that final—oh so final—chorus.

140

You're dynamite
You're outta sight
Love missile with a
Max! I! Mum! Payload!
You're dynamite
You're outta sight
Feels like you're gonna
Make! My! Head! Explode!

II
This Insubstantial Pageant Faded

THIS ROUGH MAGIC

The magician delicately places a piece of coal on a brass cradle atop a waist-high wooden stand just in front of the crushed—and, Zal can't help but notice, frayed and fading—purple velvet curtains at the rear of the stage. There's a pillar either side of the drapes, creating an ad-hoc proscenium arch, though their purpose is purely to support the deck above.

'As you can see, merely an ordinary piece of coal,' the magician says. Like his drapes, his affected aristocratic accent shows the effects of wear, small gaps occasionally revealing the less refined material immediately beneath. 'But given enough time, and pressure, it will turn into a diamond. I have to confess I've been waiting a while, and up on this stage here I can certainly feel the pressure, but there's still no sign of progress. We live in hope, nonetheless.'

His patter is greeted with a few laughs out of politeness and sympathy, bordering upon embarrassment. Zal doesn't offer any fig-leaf chuckles of his own; it would just be too patronising, and the guy deserves more dignity than that. Any old stager does, even—maybe especially—when their gifts are long since faded and the final curtain is unavoidably beckoning. Under the stage lights, The Great Mysto, as the sign in the lobby advertises him, looks well into his sixties despite the make-up. He's a slightly built man who must have been dartingly nimble once upon a time, a quickness on his feet matching a quickness of eye and mind. Now, though, he seems tired, his motions too deliberate, his stagecraft clunky and functionary, no spark to what he's about. He looks defeated. His assistant knows it, too. She's far younger than him, maybe early thirties, slim, petite and lithe as the position dictates, and wearing less make-up than he is. She keeps smiling as she faces the audience, but she looks tense half the time, afraid something's going to go wrong at any second. With saws and swords lined up as part of the repertoire, Zal muses, this is an understandable anxiety.

The venue is barely a third full, and it's not a big room: half the size of the suite next door where a stand-up comic is billed to appear later, and a quarter the size of the ship's ballroom. The audience comprises mostly families, but half the kids aren't even paying attention. It being term-time for school-age children, it's mostly toddlers, who are just wandering around in the aisles. There are a few older ones, uniformly playing with their Gameboys. The adults are just as fidgety and inattentive, impatient for the show to finish so they can pack the kids off to bed and finally hit the bar, maybe even come back here to see a 'proper' show later. This is the early bill, scheduled partly to suit a family audience, and partly so that the room is free for the jazz/blues act billed to play the late-evening session.

'Pressure,' he says, as if trying to reboard his own train of thought before it chugs off without him. 'Pressure: that's the applied principle behind the sword.' As he says this, he pulls a long blade from a rack, while his assistant slides a wicker basket from the side towards the centre of the

146

stage. 'The sharpness of the steel is essentially a means of concentrating pressure on the smallest possible area, allowing even a frail little man like me to defend myself with the minimum effort.' At this, the assistant reaches into the basket and produces an armful of fruit, which she throws at him with impressive rapidity. The Great Mysto merely holds the sword vertically in front of him, the hilt gripped at waist-height, and the missiles are bisected: two limes, two lemons, two oranges, two grapefruit. This does raise a few chuckles, the fruit getting larger and the assistant's apparent ire greater as she escalates her unilateral arms race. This culminates in her producing two weighty-looking water melons from the basket while the magician, apparently oblivious, turns face-on and gives a little bow as though the trick is finished. The assistant then lobs them, one from each hand, in a one-two movement so that both are in the air simultaneously, in response to which the Great Mysto nonchalantly flicks his wrist and angles the sword so that both melons are impaled upon it.

The assistant looks a little sheepish and the magician gestures to her to climb inside the erstwhile fruit receptacle. She complies with an exaggeratedly huffy pout, whereupon the Great Mysto commences the standard swords-through-the-basket routine.

Mysto's not without his moments, but there's something wrong, and Zal has worked out what it is. It took a while to occur to him, but then only a little more observation to confirm. There were no coins, no cards. The guy isn't doing any intimate stuff, no close-up work, which to Zal seems an even more conspicuous omission given the inescapable modesty of the venue. A little room like this, a small audience, and the guy hasn't so much as fanned a deck. Even the magicians playing theatres in Vegas would start off with something simple to set the scene, help to pace the ascent to grander illusions. Why wouldn't a veteran conjuror playing under these reduced circumstances not get the little kiddies onside with some coin work, a few vanishes and transpositions before paying off the little bastards with a few miraculously appearing bribes?

As soon as Zal asks himself the question, he realises his query already contains the answer—veteran conjuror—and a little scrutiny verifies his hypothesis. It's his hands: the poor sonofabitch has rheumatoid arthritis, and has constructed an act out of what illusions his stiff and irreversibly deforming fingers can still execute, with an inevitable weighting towards machinery: self-working tricks and automated gizmos that ought really to be used more sparingly to augment a wider repertoire.

Zal feels it right in the heart, pictures the slow deterioration, the moment the guy tried a particular palm or sleight and realised he could no longer manage it. What moves him all the more is that the guy is still clinging on to flotsam long after his ship has been wrecked. He's been unable to give it up and is still performing his failing act before dwindling and disinterested audiences aboard a cruise liner. Zal can't decide what would be the sadder: if he is doing it because he still needs the dough, or doing it simply because it is all he knows. Probably a combination of both.

It sure helps create a breathless tension to

Mysto's execution of the old basket routine, but unfortunately only for Zal, who can't take his eyes off the old man's hands as he thrusts each sword through tiny slits in the wickerwork. If he had played up an image of bumbling clumsiness, it would perhaps have engaged the audience more, but the overture to the trick had been about precision, emphasising not only the sharpness of the blade but the control with which Mysto wielded it. Thus the audience were more taken with the overture than the trick itself, the former having the novelty value of an element of the unexpected and unpredictable. The latter is something they've seen a hundred times on TV, and even seeing it live makes no difference, because Mysto just doesn't have the stagecraft to get them excited about it. Zal's witnessed comics perform classic gags the audience have already heard over and over on TV, but they know how to play the material so that they still get the laugh: they know how to *sell* it. Mysto must have known once, but now he's just going through the motions, his audiences' indifference having created a kind of positive-feedback loop so that he cares just a little less with each performance. Thus the paltry crowd today has little interest in the outcome, and even less appreciation of the skill and ingenuity being practised by the two performers. They know the girl is going to climb out of the basket unharmed, so every sword he drives in is one more he'll have to pull out again before the trick is concluded: meaning each blade is not one sword more than could explicably go through that basket without injuring the girl, it's one sword more between them and that drink once the kids have been safely tucked up in their cabins.

Zal, the Great Mysto and the weary parents are all aboard the cruise ship *Spirit of Athene*, three days out of Palma and currently en route to the Canary Islands via Casablanca and Agadir. It had put in at Palma for forty-eight hours and, as it turned out, was due to depart at exactly the same time as Zal's flight to Paris. He didn't believe in fate or providence, but sometimes you just took your cues from the signs life randomly threw up. What else was he going to do: throw a dart at a map? Head for the airport and play departure-board roulette? A cruise liner was not exactly his scene, but it seemed an effective way of dropping off the screen at zero notice. It was also a concentrated version of the lifestyle he'd adopted for much of the past year: the ultimate in travelling hopefully but never arriving.

Bitch of a place for anyone to abduct him from, too. Bearing that consideration in mind, he'd booked his berth online at an internet café, using Albert Fleet's credit card details to pay for it. He'd got all he needed from the wallet, including the guy's mailing address from his driver's licence. That was partly why he'd made play of taking the hard cash and dropping the wallet itself: make him think he'd no interest in the other goodies. Fleet wouldn't know anything was wrong until he got his next bill—not unless they flagged it up for authorisation, but Zal figured spontaneous travel fares weren't exactly a conspicuous rarity on this guy's monthly statement. The other way he might find out sooner would be if Zal went crazy and maxed out his credit limit, but there was no call for that. He didn't need the guy's money, just needed

150

him to pay for his passage. It was appropriate, after all: Fleet had been planning to set up Zal with a cosy cabin aboard a boat leaving Palma that morning anyway, hadn't he?

He signed himself up for a month, due to finish up at Marbella on the return leg. Time enough to reflect and consider what he was going to do with a future that he had now resolved would not, could not contain Angelique.

The second day out, at noon, he marked the moment when he was scheduled to meet Angelique at the Musée d'Orsay by leaning against a guardrail and firing cards into the water with practised flicks of his wrist. He tried not to wonder whether she'd even be there, tried not to picture the scene. When he got to the last card, he turned it over to reveal it as the eight of diamonds: the card she'd chosen (or rather the card he'd forced on her) once upon a time at the Louvre. This was no fateful coincidence, however, as he had earlier flicked through the deck, selected it and slipped it to the bottom in a sequence of moves now so natural and automatic as to be almost subconscious. He looked at it for a second, thought of keeping it as a symbol, but a symbol of what? Some token he'd sometimes lie awake and fantasise about giving her should they ever meet again? No, *here* was a symbol: let her go.

He threw it and watched it spin on the breeze, losing sight of it against the bright blue sky.

Farewell.

His third night at sea, he'd had enough of his own company, fed up soul-searching and just about done feeling sorry for himself. He decided to check out the onboard entertainment. It would be a waste not to: Albert Fleet had worked hard to earn the money that was paying for this, after all.

The assistant has retreated offstage, leaving the Great Mysto to run through a few tricks on his own. Watching him work—clunky and struggling—Zal is feeling a strange disquiet, a restlessness of spirit. It's growing like a goddamn itch. He observes Mysto go through the motions and can't help but remember sitting in a lounge in Vegas, watching his dad do all of this so much better: so, so much better. Even hungover, even half drunk, his father was several classes above this shambles, and probably several classes above Mysto even when the latter had been at the top of his game. He remembers the view from the side of the stage, once he was a little older, after Mom died and Dad straightened out, when his old man employed him behind the scenes. He feels the restlessness turn into a twisted longing that churns him up inside. He's recalling the conflict within, the price he paid for a denial borne out of spite and blame and anger and grief and all the things a boy his age could neither comprehend nor contain.

'This stuff can be taught, son,' his dad stated. '*I* was taught, and I had a great teacher. But you've got something natural, something that cannae be taught. You could be among the best, son, way above my league.'

Time and again his dad would tell him this, and that angry, hollow part of him enjoyed hearing it because he knew how much it hurt the old man. Not becoming a magician, not realising his own potential and his father's dreams, that was his revenge for how the old man had failed him and Mom, his payback for the neglect and selfishness

152

of the alcoholism that caused Zal to blame his dad for her death as much as the asshole drunk-driver who killed her.

Yeah, he really socked it to him with that shit. If you really want to hurt someone, hurt the one he loves. Zal's revenge had been an ongoing act of self-harm.

Seeing the not-so-Great Mysto right now, it all comes back, though the memory of his anger is transmuted into sadness and regret, and it is the feeling his anger was in conflict with that comes through strongest. He felt it when he was seven years old, he felt it in his teens, he felt it even as he told his dad he wanted nothing to do with his profession, and he felt it every night he pulled a crowd in the Dracon Rojo.

The Great Mysto completes a passable cup-and-balls routine. Zal watches his thumb and forefinger, his wrist action. The appropriate joints are still supple enough to turn the moves, though there's an awkwardness about his grip that would be jarring to a more attentive audience. Zal flexes his own digits, feels the absence of a coin like it's the absence of a finger, and reaches to his pocket to remedy this. He flips it, turns it, passes it, repeating the moves absently like a nervous habit as the show limps on.

Mysto procures a volunteer, picking on a parent near the front, knowing an adult will be too polite to refuse. Zal wonders what he's going to attempt, and is set on edge to see the magician break out a deck of cards. The woman briskly approaches the stage with the same combination of the dutiful and sheepish, as though she was retrieving her errant toddler from having strayed into someone else's cabin. Mysto executes a three-quarter circular fan of the deck, which looks graceful enough in motion, but fairly ragged and uneven by the time he's presenting it to his 'volunteer'. He offers her the fan and asks her to pick a card, which she must show to the audience but not to him. It's the ace of diamonds; Zal's guessing a force, given it's a big card, wonders if it might even be a forcing pack.

He fumbles slightly as he closes the fan, inviting her to place the card where he has cut the deck. Zal wonders for a second whether it's a mask for his technique, the moment of clumsiness creating the offbeat in which to execute a switch, but no: this isn't a trick, it's a tragedy. The legendary Cardini feigned being drunk, but that was an intrinsic part of his act, a running gag between him and his audience and thus part of the *charm* of his act. This, by contrast, is merely painful to behold, and of the audience, only Zal is even capable of noticing.

The lady scuttles hurriedly back to her seat, while Mysto walks to the front of the stage, cutting and recutting the pack.

'In order to find your card, madam, I'm going to perform a special shuffle. It's one that takes years to perfect, so don't anybody ask me to teach them it afterwards, and don't ask me to repeat it, because at my age I'm lucky if I can get it right once. Okay, here goes.'

Zal leans forward on his seat, the coin between his fingers suddenly held still in suspense. This is even worse than the basket illusion. Mysto cuts the deck once more, taking a half in each of his gnarly and slightly malformed hands. Though he feels like he ought to be watching through his fingers, Zal's eyes are locked on to Mysto's own digits, his thumbs springing tension into the cards, preparing a horizontal spring riffle.

The magician's hands suddenly spasm and the cards explode from the collapsing cradle of his fingers, spraying, spinning, fluttering about the stage like crisp autumn leaves stirred from the gutter by a sudden gust. It is not a flourish, but a fumble, a moment of startlement. A trick derailed, an unscripted incompetence. Some members of the audience gasp, others fail to stifle giggles. The muted laughter is horrible: a cringing combination of being embarrassed on the faltering magician's behalf and being embarrassed by being present at such a tawdry spectacle. But can he recover, that's the question? Does he have an out?

The Great Mysto looks up once the last of the cards has fluttered to the ground, then gives a grin.

'Told you I can only do it once,' he says, eliciting a couple of sympathetic laughs. 'But I did also say it was magic. What was your card, madam?'

'The ace of diamonds,' the lady replies.

The magician turns side-on and gestures to the rear of the stage. Sitting on the brass cradle, where the lump of coal was resting the last time anyone cared to look, is the card she named.

'And what do you know, it turned into a diamond in the end.'

The magician takes his bow, joined shortly by his assistant. There is applause, mostly out of relief that the show is over. Zal, however, offers his more sincerely. The Great Mysto might be failing and ailing, but kudos to him, kudos indeed, as he really drew Zal with that last trick. He liked everything about it, everything it represented. Zal guessed the

155

assistant had reached through the curtain and performed the switch during the disastrous spring riffle, which was poetry: his revenge on his debilitation, and the only moment he'd shown a genuine twinkle in his eye.

The house lights come up and the audience begin to bail before Mysto and his assistant have even left the stage. Not everybody makes for the exit, however: Zal spots a man in a suit, with a laminate clipped to his breast pocket, standing just inside the door, leaning against the wall, arms folded. Having had his eyes on the stage, Zal doesn't know how long he's been standing there or how much of the show he's seen, but from his expression, figures an accurate estimate would be 'enough'. He waits until the performers have retreated from sight and begins making his way towards the front, treading on to the stage and through the curtains with the unmistakable proprietary air of officialdom.

Zal remains in his seat, the only spectator left as a couple of cleaners enter the room with a cart and begin moving through the rows clearing trash. He figures the popcorn boxes and candy wrappers aren't the only things getting shit-canned right now.

The guy in the suit emerges only a few minutes later, pausing for a moment to let out a long sigh before heading for the main door. He doesn't look like he enjoyed what he just did, so give him that much.

Serendipity: is it being in the right place at the right time, or are we frequently in the right place at the right time and serendipity is the name we give to those rare moments when we have the vision to realise it?

Zal waits for the guy in the suit to leave, then makes his way on to the stage and slips behind the curtain. The area beyond is deceptively large, far deeper than the view from the front suggested. The Great Mysto has draped it to control the sightlines, bearing in mind the two pillars, and delineated a small performing area, leaving an unused space at least three times the size of what the audience can see. For this reason, neither he nor the assistant notice Zal slip through the velvet and stand still, taking in the plethora of props and equipment ranged untidily about the place; some of it deposited haphazardly during the performance, other items veteran remnants of shows gone by: artefacts of the Great Mysto's magical history, uncatalogued museum pieces, impeccably preserved but gathering dust.

Mysto and his assistant are a good ten yards away, the magician sitting on a closed trunk, his back to Zal, blocking him from the girl's line of sight. Zal figures he could be right next to them and they'd not notice, as they seem isolated in their own intimate little world of despond. Zal hears the man cry, sees his shoulders shake with a sob, sees the girl's arms move around him, the top of her head visible as she leans over and presses her face against his chest. Zal hears her sniff.

'Oh, please don't cry, Dad, you're setting me off now.'

'I'm sorry, love. I'll be fine in a minute. Don't know why I'm like this: I knew what was coming, we both did. It's just . . . well, I'd like to have known it was my last show *before*, you know? In advance. Not find out like this that it's already

157

gone. Best part of forty years and it all ends just like that, with a whimper. My bloody whimper,' he adds with a bitter laugh.

'Just remember that the best part of those forty years wasn't on this bloody tub, Dad.'

'Aye, that I know, Lizzie. And it's a mercy, really. You ought to be spreading your wings, girl. Not stuck here propping up an old fossil.'

'Don't be daft, Dad. Besides, nobody's been battering down my door, either.'

'Well, more fool them. Ach, I should have given up ages ago, that's all I'm saying. Never dragged you along after I ought to have realised my time was up.'

Lizzie sniffs again. 'I wouldn't have missed it, Dad,' she says.

Zal is conscious of his apparent invisibility and decides he ought to draw attention to his presence before he feels any more voyeuristic. He has been inspecting the contents of a well-travelled but formidably constructed wood-and-leather trunk, so lets the lid drop closed. The bang is cushioned, testament to the craftsmanship, but audible enough for Lizzie to stand up straight and look over her father's now slowly turning head.

'Can we help you?' she asks, with an aggression that instead states: 'Can we help you rapidly fuck off out of our privacy?'

Zal understands what he's just walked into and knows he has to defuse it. He also knows that simply being here and derailing their grief is already half the battle. He now just needs a little of the conjuror's stock-in-trade: misdirection.

'Yeah,' he says, bending down briefly and passing an appreciative hand over a device on the

158

floor, rolled back there and discarded during the performance. 'This flower cascade looks like a Bautier deKolta. It's not . . . surely not an original, is it? And if my eyes don't deceive me, that looks very much like a cocktail bar set-up after Alan Wakeling.'

Lizzie comes stomping towards him, wiping a tear but indignation thoroughly erasing the grief from her expression.

'I don't believe it,' she mutters to her father. 'That tosser Henderson tipped the wink to some bloody collector before he even informed you. Of all the disrespect . . .'

'I'm not a collector,' Zal informs her calmly.

'Wasting your time if you were,' says Mysto, getting off the trunk he was sitting on and turning towards Zal. 'They're neither of them originals. Made 'em meself, but you could call them reproductions.'

'I'm betting you made the trunks too. The deKolta is first class, great to see something like that still in action.'

The old man can't help but smile, as much at a shared appreciation of the golden-age apparatus as at Zal's compliment to his craftsmanship. Lizzie stands vigilantly with her arms folded, like a bouncer waiting for the nod to kick Zal out into the street.

'I did have first-hand access to the originals,' Mysto says. 'Collector got me to manufacture a working copy of the cascade so he had a functional model as well as one actually crafted by the man himself, and I just made two. For the Wakeling copy, I had to procure a private examination.' He raises his eyebrows involuntarily at this, the memory of a doubtless dubious undertaking. 'Haven't used it in a while, though. It's my hands, you see.'

Zal nods.

'Bet it worked beautifully back in the day, though.'

'Oh, what? Brought the house down,' he says with a sad smile, which turns that bit brighter and prouder as he adds: 'Though not as much as my deKolta Expanding Die.'

'You gotta be kidding me. Jesus, didn't even *Houdini* covet the secret of that trick? And you made one yourself?'

'Houdini paid Goldston a fortune for it, yes, but then Goldston published the plans so Houdini wouldn't have it exclusively, one of the few times a friend or a rival—and Goldston was both—put one over on him. Happily for the likes of me, the design's long since been out there for those as have the know-how and the skill to give it life.'

Mysto warms to his subject, already talking to Zal with an openness and enthusiasm that seems to have forgotten their circumstances. Lizzie hasn't, and decides to intervene.

'I really am sorry, Mr, eh . . . ?'

Zal turns to look at her, takes a beat. 'McMillan,' he tells her.

'Mr McMillan. This isn't the best time right now. Sorry to be brusque, but can you tell us your business, please? If you're not a collector, then who are you?'

Jolted out of reverie by his daughter, Mysto remembers himself enough to be suspicious rather than merely curious. 'Aye,' he says inquiringly. 'What kind of man can recognise a deKolta, never

160

mind an Alan Wakeling reproduction?'

'The kind of man who saw Alan Wakeling use it live on stage, though I was very young, so I remember it better from seeing Earl Nelson perform later using Wakeling's original rig. I'm a man very much like yourself, Mr . . . ?'

'Morrit. Daniel Morrit. Like me how? A magician?'

'An out-of-work magician. One could in fact say, very accurately, that we're in the same boat.'

'If that's the standard of your patter, no wonder you're out of work,' Morrit says grimly.

'So you *were* tipped off,' Lizzie observes.

'No, I'm just a keen observer. Keen enough to spot that the Great Mysto can't handle coins or cards any more, though not keen enough to spot you switching that ace for the coal.' Zal smiles. 'I liked that.'

Lizzie's arms remain folded but there's a hint of pride in her face at this compliment. Zal can tell her defences are lowering, just a little.

'I want to make you an offer,' he says.

'For what?' Morrit asks grumpily. 'The deKolta?'

'No, for everything.'

'Job lot? I don't know, son. Some of this stuff I'd never sell, other of it's not worth a wet fart. Besides, it's too soon for me to be thinking about it. For God's sake, I only just—'

'When I say everything, I don't mean your stuff. I mean the show: like a franchise. Or more like a partnership. I'll confess, watching the performance, I was just planning to offer to buy you out, but when I came backstage and saw some of this stuff, I realised . . .'

'Hang on a minute, son,' Morrit says, screwing his face up like he's trying to do trigonometry in a hurricane. 'I think you're forgetting what happened *between* you watching the show and you coming back here. There *is* no show. We've had our cards, remember? Henderson's paid us off. Settled the rest of our contract but told us we're not performing any more. Said we were bringing down the tone, not up to the class they want the ship associated with, and he's not bloody wrong, either.'

Zal nods, unperturbed.

'Were you contracted for just this cruise, or a month, six months, what?' he asks.

'Just this cruise, but it doesn't—'

'That's ideal. Means we can negotiate a better contract all the sooner. That's assuming you and I can reach an accommodation. I'd need you to be my technical consultant as well as my craftsman, and I'm rusty on a few techniques so you'd be my tutor too. You'd retain ownership of all the materials but we'd be fifty-fifty on any new apparatus our collaboration happens to conceive.'

Morrit starts shaking his head. The trigonometry face has loosened and given way to an expression of patient amusement at the ramblings of a fool. He steals a glance across at his daughter, whose incredulity looks neither patient nor amused.

'What part of "we're fired" are you having trouble grasping, Mr McMillan?' she asks.

'All of it,' Zal shoots back, eyeing Lizzie with his own impatient sincerity. 'Look, I appreciate that this is the bit in the musical where it looks like the show won't go on, but believe me, it will. Your

162

show is over, that part's true, but this guy Henderson has a boatload of people to entertain, and if we stage a new show that puts asses on seats, that's all he's gonna care about.'

Morrit's starting to get it. 'A new magician,' he says, thoughtfully.

'Exactly right. Now, the ship's putting in at Casablanca for two nights. I figure we can be ready by the time she sails again. Henderson sees we've got a big crowd, that changes everything. Weather forecast for the next week is not good—lot of cloud, lot of rain, lot of people stuck indoors on this hulk. Word gets around, pretty soon they'll be blowing the cobwebs off the House Full sign, you wait and see. I'm betting by the time we hit the return leg after the Canaries, we'll be in the big suite next door, which is just as well, because it's got a trapdoor and this room doesn't, and I wanna start with the Expanding Die. You up for that, Mr Morrit?'

Morrit's growing grin heralds his answer. 'Bloody right I am. What have I got to lose?' he asks Lizzie, whose face is urging caution, afraid an already fragile man is getting carried away on false hopes.

'You talk a good show, Mr McMillan, I'll grant you that,' she says. 'But it begs the question, if you're that good, how come you're out of work?'

Zal thinks of the countless hours he spent practising in jail. He thinks of robbing a bank in broad daylight and in front of a legion of cops; of what he pulled off later under the noses of just as many police and a few dozen gangsters. He thinks of how amazing his dad was onstage, and he thinks of his dad's words, the ones he always discarded or

163

threw back in his face: that he had a natural gift for reading the audience, that he had timing, grace and touch 'like I can only dream aboot, son'.

'Let's say I had a sabbatical,' Zal replies. 'But yeah, I'm pretty good.'

'You'd better be,' she says, extending a hand for him to grip, the welcome sign of her assent. 'Especially as your first trick will be selling us back to the man who just dumped us.'

'Oh, I'm not going to tell Henderson a thing. The first he's going to know about it is when he comes to investigate why there's a packed house in a lounge where nothing's scheduled. We're going to present him with a fait accompli.'

'But how are we going to get a crowd?' Morrit asks. 'I've every fear I've already stifled any appetite there was for a magic show on board this ship, and if we're doing it hush-hush and without official sanction, how are we going to advertise the bloody thing?'

'That's *why* we'll pull a crowd,' Zal tells him. 'Because we're gonna advertise our show the old-fashioned way.'

Morrit's eyes narrow conspiratorially.

'I like your style, Mr McMillan. Tell me, what should we call you, professionally?'

Zal pauses a moment, gives a smile to suggest he's taking a beat to build anticipation. Truth is, he hadn't thought. First thing that comes into his head is his dad's stage name, which he can't use for any number of reasons. Second thing is what his mum used to call him sometimes, her Mexican playful corruption of McMillan, the middle name his dad gave him.

'Maximilian,' he says.

164

Morrit nods approvingly, but Lizzie still seems perturbed.

'You don't like it?' Zal asks her.

'I like it fine. Just . . . forgive me: I dropped out of a business degree, so maybe I didn't learn enough marketing and advertising theory. Can you tell me what the old-fashioned way entails?'

*　　　*　　　*

The *Spirit of Athene* is three hours out of Casablanca and sailing in heavy rain, as forecast. The larger of the ship's indoor swimming lidos is extremely busy, not a spare plastic chair to be had, never mind a reclining lounger. Families have set up camp for the day, parents sitting amid piles of toys and towels. It hasn't been warm enough for any but the most dedicated swimmers to brave the outdoor pools on this cruise, but many of the older passengers have been content to sit by the sides reading their papers and books, wrapped up in jackets and sweaters. Strangely, they always face inwards around the pools rather than out towards the ocean, almost as though they're just waiting to finish one more page before they suddenly strip off and dive right in. Today, though, the decks are awash, but the hypnotic draw of chlorinated water has endured, and thus dozens of those with no intention of taking a dip have opted for the indoor poolside as their preferred spot to pass the rainy day.

Two men dressed in black dinner suits enter the lido carrying a large trunk between them, preceded by a petite and shapely woman in the eye-catching if impractical combination of a gold bathing costume and matching high heels. She sashays in like the poolside is a catwalk, one arm held aloft and her palm bearing a pair of handcuffs which dangle around her wrist. She's smiling, making eye-contact with as many observers as possible as she proceeds around to the deep end, where the two men stop and deposit the trunk about a foot from the edge of the water.

While the two men lift the lid of the trunk and gently swing it open on its hinge, the woman skirts the row of sunloungers nearest the deep end, scanning the observers. She sees a likely candidate and extends her free hand, inviting a bashfully grinning bloke to take hold of it. No sooner have his fingers touched hers than she has brought around her other hand and cuffed herself to him. She walks away, forcing him to his feet and leading him across to the trunk amid laughter, whistles and cheers. One of the men by the trunk proffers a set of keys, which the woman takes, using them to free herself and then cuffing both the grinning man's hands together. His daughter, a little girl of about three, runs up and hugs his legs with a look of confused concern. This elicits further laughter, then an appreciative 'aaaw', better than anything they could have scripted, as the woman surrenders the keys to the little girl in order to free her daddy. The man takes the keys from his daughter and unlocks himself, to much applause from all around the pool. He offers the cuffs back to the woman, who takes them from him and presents her cheek for a kiss, which he blushingly delivers. Then he walks away, or rather attempts to, before realising she has cuffed herself to him again.

The bloke sportingly endures another bout of

laughter before being freed a second time, and is about to make his way back to his sunlounger when the woman beckons him towards the trunk, where the younger of its bearers is undressing from his DJ down to a pair of black swimming shorts. He has close-cropped blond hair and, denuded of his clothes, reveals himself to be athletically built and heavily tattooed. The older man then enlists the help of the new recruit to remove a number of items from the trunk: a length of heavy chain, two formidable-looking padlocks and a beige sack with aluminium eyelets puncturing its neck. The recruit is then invited to fasten the handcuffs behind the tattooed man's back. The tattooed man slowly rotates himself three hundred and sixty degrees in order that everybody around the pool can see how thoroughly he is restrained.

The older man then opens the sack and the tattooed man steps into it. He crouches so that the neck can be pulled all the way over his head, then the heavy chain is passed through the eyelets and the sack pulled closed. The recruit is directed to lift one of the padlocks and uses it to secure the chain. Next, the two of them take hold of the man in the sack and lift him into the trunk. Once he has been placed inside, the lid is pulled closed and the recruit invited to attach the second padlock to the hasp on the trunk's front.

Finally, the recruit is asked to take hold of one of the handles on the other side and help lift the trunk off the ground. He looks extremely unsure of himself as the older man indicates that the next task is to drop the trunk into the swimming pool, but the woman is already priming the crowd to count one-two-three in time with their swings.

167

The crowd counts, the pair lets go on three, and the trunk splashes into the water where it sinks swiftly to the bottom, air bubbling almost angrily from it for a few seconds after it comes to rest.

Many of the crowd get to their feet with the intention of getting a closer look, but the woman warns them to stay back from the pool. They obey, automatically investing the woman with imperative authority. Maybe it's the heels, or more likely the need for reassurance that somebody is in charge of this and by extension that all of those involved know what they're doing.

The woman asks the crowd to hold their breath. It's only when she says this that some of them realise they've done so already. She tells them that it's the best way of knowing when they ought to think about intervention. A minute passes. Two minutes. There are exhalations all around the lido as vital capacities are tested and found wanting. Laughter turns to silence as each panting proof of failure further builds the tension. The surface of the pool is calm, no movement disturbing its slow re-establishment of equilibrium since the submersion of the trunk.

Unseen inside, the tattooed man is checking his watch, for timing is everything. The quintessence of every escape trick is that it should seem last-gasp, thus he must let the tension grow and the seeds of fear be sown, but he should not stretch credulity nor scare the audience so much as to render the tone macabre or tasteless. There's air enough for a few minutes more, but it's also crucial to remember that a box at the bottom of a swimming pool is not a very interesting sight.

He was free of the cuffs even as the sack was

being pulled over his head, free of the sack before the trunk was lifted off the tiles. He had to be, in order to press the switch releasing the air that generated all those bubbles, so vital to conveying the impression that the trunk had flooded.

He decides it's time. He engages the concealed hinges on the padlocked side where the lid meets the trunk, then slides the switch that releases the hinges on the opposite wall. A few seconds later, he is climbing out of the pool, each of his hands then gripped by one of his companions, before all three of them take a bow, enthusiastic applause echoing all around the lido.

He announces to the gathering that if they wish to see more, he will be performing that same afternoon, with the details to be found on the cards his two companions are now passing out to a multitude of eagerly extended hands.

'And that,' he tells the woman in the gold bathing suit, 'is the old-fashioned way.'

NOT SAFE FOR WORK

Angelique is standing next to the stage inside the Tivoli nightclub, where the TV crews are starting to dismantle their equipment and the last of the hacks are drifting away. She finds there's something unsettling about being in a nightclub during the day, with the house lights up and the music off. It's not the disappointment of being behind the scenes, of an illusion dispelled, to which the access-all-areas pass of her police warrant card has often exposed her. There is undoubtedly something that kills your dreams, that extinguishes an enduring ember of childhood wonder when her investigations take her backstage at the theatre, or through the sets of a theme-park ride with the machinery stopped and the lights on full, but that's not the sensation here this morning. Instead, far from dispelling the illusion, it seems all the easier to imagine the place in full swing, to hear the music, see the gels and strobes, the bottles and glasses, the pulsating throng on the dancefloor and the tentative couplings discreetly and palpitatingly progressing in a dozen dusky recesses.

What's gnawing at Angelique is a sense of missing out: an awareness of other people's good times she can't be a part of, and not merely another grass-is-greener self-torturing fantasy. It's a memory, a recall of the way it felt to be inside a place like this, twenty years and fifty corpses ago. She remembers the promise of such places, as thick in the air as the perfume, pheromones and cigarette smoke: the tantalising possibility that tonight you might meet someone, and it would be perfect. It was a promise from the time before knowledge and experience could corrupt the fantasy, before her imagination could vividly exhibit in advance all the ways in which any potential relationship wouldn't work out.

The press conference is long since wrapped up, but she's waiting here for someone. She'll be glad when the last of the media shower have bailed and the cops have the place to themselves again, as she's been feeling understandably self-conscious. When it's been an agreed strategy to make sure the cameras and the eyes of everyone present are fixed upon you, it's hard not to feel a little on-the-spot. Her moment under the limelight, the orchestrated part deliberately intended to make her the focus of so much attention: that was the easy bit. Ironically, it was like being undercover, a paradoxical sense of refuge and anonymity to be found in the act of being in character, playing a part. What she wasn't used to was this new experience of not being able to exit the stage and take the mask off. Growing up in small-town Scotland with a brown face had given her plenty of early experience of what it felt like to be stared at, but there was never any chance of becoming inured to it, and happily in adulthood it had seldom been a problem. Today, though, there was no escaping this curious scrutiny, the awareness of remaining the focus of curious glances long after the spotlight was back on Detective Superintendent Dale during the press conference, or indeed after the whole show was over. It came with the unspoken words: 'That's her', and an unsettling awareness that people felt such casual spectating was their right, that she had

171

somehow just become public property.

Even more vertiginous was the inescapable sense of certain unstoppable processes having been initiated, over which she would now have little if any control. A juggernaut with no brakes had just rolled over the brow of a hill, and she was not so much being handed the wheel as being strapped to the front and used as a hood ornament. On the upside, if she really was serious about getting out of the police, then this was, at least, an irreversible step towards the exit in as much as it precluded any future undercover work. Taking away the thing she did best and felt most comfortable with wasn't going to leave much reason to stay. Maybe that was one of the reasons she agreed to come over here. That and the opportunity to absent herself from Dougnac and the drip-drip war of attrition he habitually conducted against her resolve at times like this.

There was also the trifling matter of a mass murderer she had thought long dead apparently still walking the earth and practising his forte with renewed alacrity.

Shaw, it turned out, wasn't part of the investigation. He had just been the conduit, a friendly voice on the other end of the phone to keep her from hanging up. David Dale was running the job, largely on the back of his success in several high-profile, publicity-intense cases, including breaking the Stockbroker Belt kidnap ring last year and nailing Marjorie Petitjean for offing both of her husband's parents, Reginald and Harriet, also known as Lord and Lady Lambton. Crime affecting rich people always got a lot more play in the media, and in such cases the ability to deal with

172

that was just as important as your skill at handling the investigation itself. Dale had kept his nerve most impressively in the face of fierce public scrutiny (as well as misinformed but unrestrainedly mouthy press criticism) throughout the kidnapring investigation; in particular with regard to suppressing information that might have sold a lot of newspapers had the press known it, though only at the unacceptable expense of further endangering the life of the ten-year-old girl Dale's men were eventually to rescue. However, as well as having the fortitude to keep the hacks at bay when he deemed it necessary, he had also played a virtuoso hand in inviting them all over the Lambton case, firing up the arclights until the 'distraught' aspiring heiress wilted under their glare.

According to Jock Shaw, it had been asserted at the most senior levels as 'imperative' that Angelique be brought into the investigation, but it had been Dale's idea to make her involvement so demonstrably public. It had also been Dale's idea to have the press conference here at the Tivoli. She was unconvinced of the thinking behind this, but her doubts about that were nothing compared to the reservations she was harbouring over the wisdom of his parading her like London Met FC's new star signing.

Nonetheless, she had agreed to it, partly because she couldn't argue with Dale's record, and more significantly because Shaw had vouched for him. She knew Shaw well enough to tell the difference between him endorsing somebody because it was professionally dutiful to do so, and genuinely anointing Dale as worthy of his own—

and therefore Angelique's—trust. It was a measure of her esteem for Shaw more than a leap of faith in Dale that she was finally assenting to a role she had been resisting for most of her police career. It was also perhaps another sign that said career was finally gleaming in its twilight.

Angelique had fought to keep herself off the front pages after Dubh Ardrain, when the police were desperate to use her as their good-news face, and had spent many of her years in the force fending off various inventive attempts to make her their ethnic recruitment poster-child. She was never the face at the press conference; she preferred being the hands on the collar and the body beneath the armour.

From here on in, however, her job would be to draw attention, like drawing fire in a gunfight. It wasn't exactly *Big Brother*: her name would be merely a caption on a few TV news reports, quoted in the print media's deeper coverage well down the page from the juicy stuff and only bumped up beyond that on slow days. Nor was she being played as bait; more a red rag to this particular bull. But to have a role so exposed, so deliberately out in the open, was a new and alien experience. Standing here in this nightclub, from where seven pop stars had been abducted, four of them already known to be dead, she had to wonder: who the hell chooses this for themselves? Who sets their aspirations upon a life permanently under the gaze of inquisitive strangers?

Dale's decision to host the press conference at the Tivoli was an understandable (though for Angelique, potentially ill-judged) declaration of intent to play their enemy at his own game. 'He wants it showbiz, we'll keep it showbiz,' Dale said. 'We can't afford to give the impression that we're downplaying anything, like we're ignoring an attention-seeking toddler, especially knowing what provoking a tantrum entails. Besides, playing these things out in the limelight is a risky business for him. He can't retreat to the shadows for his next move without losing face, looking like he's peaked and on the retreat.'

Reading between the lines, it sounded like Dale had reached the same grim conclusion as Angelique regarding the chances of the three remaining abductees. When he talked about the killer's next move, she knew he accepted that this part of the game was already lost. She couldn't criticise his judgment in how he chose to approach such a near-impossible situation; her doubts were largely around the wisdom of attempting to play Simon Darcourt at his own game when they had no real idea of what Darcourt's game might be; or indeed for sure if even Darcourt it was.

One indisputable benefit of Dale's thinking, however, was that it brought them back physically to one of the investigation's most important loci. Having come in a little late, Angelique was grateful for the opportunity to get a feel for the place. There was plenty of footage, naturally, and it was all being analysed pixel by pixel, but there was some fundamental polis instinct that could only be satisfied by getting a first-hand feel for places. She dealt in evidence and logic, so she wasn't talking about emotions or images metaphysically adhering to the surroundings, but she had a need to know what it felt like to be standing inside and outside of the Tivoli: walking where the victims walked, looking where the perp was looking. Seeing what he was seeing was, unfortunately, a more elusive matter.

The most analysed footage had been that from the CCTV cameras on Shatfesbury Avenue. Even with the images enhanced, all they had were a couple of fuzzy close-ups of a face wearing sunglasses (at night) and a chauffeur's cap, the shots extracted from the half-seconds of tape showing the limo pulling up and later driving away. Those were the only times that a camera was pointed face-on to his windscreen. He had needed to park in front of the Tivoli in order to block Vogue 2.2's real limo and present his own instead, but he must have taken the camera positions into account. He had parked a little back from the red carpet, his vehicle's nose just about in line with it, meaning that all the time he was sitting there, the closest camera only had a view from the rear, and the nearest facing camera had its view of the windscreen obscured by the awning. The two shots lifted from when the vehicle was in motion were useless: they could have been showing a shop-dummy in shades and a peaked cap. It had been suggested that he might actually be wearing some kind of blank plastic mask, and even that he had treated the windscreen with something in order to reflect very little light from within, but Angelique didn't consider either ruse particularly probable or even necessary.

Bottom line: he wasn't going to drop a bollock and he wasn't going to throw them any bones.

They did get a plate: a French one, which was initially considered a valuable lead until a cross-Channel trace revealed it to be the same as on the

Mercedes carrying Princess Diana and Dodi Fayed on their final high-speed journey through Paris. The registration was a fake, but it certainly had Darcourt written all over it.

There had been a veritable ant-colony of forensics personnel and SOCOs crawling all over Nick Foster's mansion in Kent. There was no evidence of a break-in or a struggle, and indeed consequently no evidence that it had even been this house that he'd been abducted from. Even attempts to narrow the window on when he had been taken had been fogged by apparently crucial information disintegrating before their eyes. The last known contact, it had been established, was a text to his PA, Susie Russell, the afternoon before the party. The police were already constructing a possible timetable of events placing this as the earliest possible starting point, when somebody had the presence of mind to take a closer look at what the text message actually said:

Off for a detox down-payment in advance of the shindig. Doing some penance upfront. Plan to get into a shocking state! Take tomorrow off. I'll get to Tivoli by alternative means. Stand by for a surprise entrance. ;)

The apparent comic prescience of the wording retrospectively suggested that Foster never composed the text, so he could have been abducted any time up to twenty hours before that, when he last actually spoke to his PA on his mobile.

In the meantime, however, the Forensics team

177

had, against all expectation, pulled out a plum. Concentrating very specifically on the shelves housing Foster's CD collection, as highlighted in Darcourt's video (as well as his computer, as suggested by the mention of his playlists), they had found a partial fingerprint that didn't match Foster's own dabs on the spine of an album. The fingerprint analysis lab had then undertaken a laborious elimination process, checking the partial against prints volunteered by Foster's friends, relatives, employees and anxious ex-lovers (anxious as in anxious to eliminate themselves as suspects). They were on to their third day without a match when one of the analysts, perhaps inspired by working late into the evening, demonstrated sufficient musical knowledge as to suggest that there might be more than incidental significance to the identity of the album that the elusive partial print had besmirched. It was *Rust Never Sleeps*, by Neil Young and Crazy Horse. Track six, understandably well known among the analyst's profession, was 'Powderfinger'.

Blind fate seldom hands you such neatly packaged anecdote material, and sheer luck seldom gifts you a solitary print at a locus around which the suspect has been vigilant in leaving no other traces. The analyst took the hint, and on more than a hunch tested the partial against a comparison set lifted from the home of a person no one had previously thought to check against, due to the normally compelling eliminating factor of being dead.

It matched.

The print had been left by Darren McDade's finger. Cross-checking confirmed that it was the middle one.

Indeed.

It was concluded as most likely that the killer brought along the CD for the purpose, complete with print, rather than taking along the finger itself. He hadn't been shy of distributing McDade's body parts, as the late pundit's former employers could attest, but by consensus it was agreed as simply too implausible that Foster would own something as musically substantial and enduring as a Neil Young album.

The disappearance of McDade hadn't yielded any worthwhile material so far either. Initially it had been believed that he was abducted from the Birmingham hotel room he was booked into for the night, after taking part in a TV show filmed at Pebble Mill. However, this theory had proven valuable only by way of illustrating the cops' suggestibility in the absence of real evidence. In the videos, McDade is seen waking up in a hotel room, and as they knew he was checked into one, they automatically began reconstructing the events as bookended by these two facts. Security images from the hotel lobby and the street outside showed McDade leaving the hotel and getting into a car later confirmed as having been sent by the production company, but there was no footage of him ever returning. Witnesses stated that after filming the programme, he proceeded to tan the Green Room for all the red wine they had and all the canapes he could physically keep down, before adjourning to a nearby pub called the Cap and Gown. He was last seen, very drunk, on the street outside the pub, apparently in search of a cab. It was now widely speculated that he may have found a limo instead.

Angelique is listening to her voicemails—three queued up from having her phone off during the press conference—when she sees him making his way hesitantly into the main body of the club. He's looking like he's expecting to be stopped and asked to explain himself, but the people he passes are too busy dismantling and removing either their own equipment or the backdrop and tables from the press conference to pay him much heed. He's in jeans and a leather jacket, small satchel slung over his shoulder, and sporting a goatee beard these days. Perhaps he thinks it makes him look more grown-up and august; she'd need to see it with the collar and tie he wore Monday to Friday. She tries to picture it. Nah. It would only serve to emphasise how his professional attire didn't quite work. You could put a suit on a geeky grebo but he'd just be a geeky grebo in a suit, as conversely so many born-in-a-suit types simply looked wrong if you bumped into them over the weekend. He is now the head of the English department at Burnbrae Academy, but he looks like he just walked out of Forbidden Planet; given that there is a branch less than a hundred yards away, chances are he has.

She waves subtly, then with a full, arm's length arc when he fails to spot her amid the ferment. He grins, making more eagerly and directly towards her.

'Angelique, hiya. Long time no see.'

'Ray. Thanks for coming. Really appreciate this.'

'Gets me out of school. Feel like I'm doggin' it but I've got a letter from my mum.'

'Flight on time? Have you come straight from

Heathrow?'

'Pretty much. I mean, I stopped in at a shop along the road there, but otherwise . . .'

'Forbidden Planet?'

He grins shyly.

'Aye. Buying for the weans these days, though. The wee yin's easy enough: Turtles and Star Wars, get it in Tesco's. But Martin's very into Captain Scarlet. And I mean old-school Captain Scarlet. Has to be the Gerry Anderson original stuff, no' the new CGI version. To think when I was twelve I gave away a Dinky Spectrum Pursuit Vehicle to my wee cousin. Bastard's probably cleared fifty quid for it on eBay.'

'Turtles and Gerry Anderson. Your kids were in nappies when I last saw them. How long was that?'

'Dunno. Five years, maybe six? You took a job in France, didn't you? When did you move to London? What brought you back?'

'I haven't moved here, not permanently. I got here two days ago. Which should answer your other question.'

He nods, his face immediately taking on a hunted look that had been only temporarily masked by their mutual pleasure at seeing each other again.

'Aye,' he says simply. She can tell he's been worried sick for days.

'When did you hear? I mean, when did you start to think it might be . . . ?'

'Two possible answers to that question. One would be when your colleague in the Glesca Polis came round to the house and asked me to fly down here, because prior to that I knew he was dead, right? The other would be when I heard about the

McDade videos. One of the kids in my class had one on his fucking mobile, would you credit it?'

'Virally spreading as intended.'

'I had my thoughts when I read transcripts of what was in the videos, but these are thoughts I've been having ever since Dubh Ardrain, you know? I would remind myself—and Kate would remind me too—that I saw him get sucked down into a whirlpool with umpteen million gallons of water when that tailrace opened. But because they never found a body . . . Once you allow the possibility into your head, you start thinking every weird act of bloodletting that happens in the world might be him. You just cannae go there. It became a running joke in our house. We'd be watching a movie, or *Lost* or whatever, and any mysterious malignant presence, one of us would say: "Maybe it's him." Like I said, cannae go there. That way madness lies. So yeah, I thought about him when I learned about Darren McDade, but as always I told myself not to give it any credence. Same deal when I heard about Nick Foster. Then your man appeared at my door yesterday. He didn't give much away, but enough to make sure I didn't sleep too well last night, fair to say. How do you know it's him?'

'First thing to say is we don't, not for absolutely sure, and we're hoping you can help us with that. But if it's not him, it's somebody who wants us to think it's him. There was a clue, one we believe we were intended to find, on the McDade hanging video. The executioner was visible in a mirror.'

'You could see his face?'

'No. He was in a hat and mask, dressed to look like a noted comic-strip malefactor late of a parish known as Calton Creek.'

Ray sighs.

'Rank Bajin,' he states. 'His calling card. How well was that known, though?'

'Sparsely. We suppressed the "Black Spirit" icon as much as possible back in his heyday: it was a valuable way of verifying that an atrocity was his handiwork and not some other nutter using it to cloud the waters. Even after Dubh Ardrain, when we thought we'd seen the last of him, we still kept it quiet, never gave that part to the press, neither the image nor what its original identity turned out to be. But still, these things leak. There were a lot of those calling cards left fluttering in the dust and rubble after his attacks, sometimes hundreds. I know of at least one occasion when the image appeared in a newspaper, but it was a Spanish one, and nobody picked up on it at the time.'

'I take it I just missed a press conference. What have you told the public?'

'The point of the press conference was not about informing the public of anything, but of informing *him* that we got the message. Apart from just the standard updates—ie we still know shag-all—the principal announcement was that I had been brought on board, officially as an expert on counter-terrorism. We know that'll get the media's wheels spinning, but the main point is to send him a signal. If it's someone just trading on his name, he won't clock the significance because he won't know who I am; but if it really is him, it bats it back to his court.'

'I have to say, it's not like Simon to leave any ambiguity about who deserves the credit.'

'That's why we're not rushing into anything. We've got to be very careful in working out what

183

we think his game is. On the one hand, he's imprinting the Black Spirit calling card on the McDade killing, and yet he's masked. He never appears fully in shot, and when he pulled up outside here in a limo, he made sure nobody saw his face. We're left asking ourselves: why would he do that if we already know what he looks like?'

'Maybe he doesn't look like that any more. Surgery, damage, whatever.'

'Yeah, and we're looking into that. Though even if he's had surgery, I'm one hundred per cent sure I'd know him if I looked the bastard in the eye. However, the other explanation remains that he's hiding his face simply because he's not Simon Darcourt. I'm hoping your input is going to help us eliminate that possibility, so that we can be more certain of what we're dealing with.'

'What do you want me to do?'

Angelique puts a hand gently on his shoulder by way of indicating that they're on the move. She leads him towards the exit, gesticulating briefly to Dale on her way past to communicate 'This is him' and 'We're off'. Dale nods. He knows where she's headed.

'We're going just round the corner,' she tells Ray. 'An audio post-production house in Soho. Firstly, we need you to have a listen to some tapes.'

They emerge into the daylight, watery spring sunshine belying a chill in the air that catches in your throat as you walk into it. It seems surprisingly bright. The lighting in the Tivoli was anything but subdued, especially with the TV gear boosting it, but she'd been in there two hours, and something about the surroundings subconsciously told her to expect darkness outside.

184

'Your man up in Glasgow left me a DVD-rom with the videos,' Ray says. 'I watched about as much as I could stomach. Audio quality isn't brilliant, but I have to say, the voice didn't sound like him at all. Weird accent, for a start, but the thing is, when I saw him at the power station, even though his accent had changed vastly from our student days, it was still unmistakably his voice. The voice on the videos is nothing like him.'

'That's why we're going to Soho,' she assures him. 'We've had some experts run the rule over the videos. They say the soundtracks have been overdubbed. They reckon he spoke in his own voice at the time, then dubbed a new, altered voice over it. Mostly they assume the new voice is saying the same thing, but there's a few points on the tape where changes in the background levels indicate what was put in didn't entirely cover what was removed.'

'But I thought the Foster thing was a live feed.'

'Everybody assumed that because it was interjected into a live event, but it was a recording.'

'So if he's overdubbing it with someone else's voice, what do you think I—'

'They don't think it's someone else's voice,' she interrupts. 'Who would you get to record something like that? Plus, fair to say we don't want to begin to contemplate the possibility that he's not alone in this deranged enterprise. No, they reckon he's digitally altered his own voice. They're trying to reverse-engineer the process, which is where you come in. They need someone who remembers his voice to give them the pointers they need to tweak the settings.'

'What, so they can make the recording sound

185

like him? But what if it isn't him? Won't that just create a huge red herring?'

'No, no. The digital effects can't put something on the tape that isn't there already. They can't record you and make you sound like James Earl Jones. If it is Simon Darcourt, then at some point, they're going to twist the right knob or tweak a certain slider just precisely the correct amount, and you're going to suddenly sit up and say: "Fuck, that's him." But if it's *not* Darcourt, then they could try every setting, every configuration . . . basically, you could be there a while, and the longer you're there, the more chance it's a negative.'

'You'll understand when I tell you I'll be hoping it's a long shift,' says Ray.

'I know. You don't need to tell me how the idea of him still being alive isn't the most welcome one.'

'We didn't really part on the best of terms. I think he might consider that I pissed on his chips a wee bit, what with wrecking his masterplan and sending him to his apparent death.'

'I know this is not the most consoling notion, Ray, but in my opinion, if he is alive and he had a mind to come for you . . .'

Ray nods with a thin smile.

'I'd already be dead. I know. Happy thoughts.'

'To which I would add that making you part of his plans was what led to his downfall last time.'

'You're ranging Simon's ability to learn from his own mistakes against his capacity to hold a grudge. That's quite a face-off. I'm aware that even though he was a bampot, he wasn't an idiot, but I also know that he was one of the most self-righteously vindictive people ever to set foot on the planet.

Not much of a one for the relativistic perspective. Once he's decided you've "disappointed" him, there's not really a lot you can do to get back into his good books. I think if you dedicated the rest of your life to some form of personal penance for whatever slight he perceived you to have committed against him, he'd still feel you owed him two more lifetimes' worth before you even got close to the forgiveness waiting list. So I'm inclined to think it can't be him, because if he was still alive, he'd have come for me by now.'

'Given the capacity for grudges and vindictiveness you just alluded to, it might be flattering yourself to think you're a priority. Maybe he's got a long roster to get through, and you'll have to wait your turn.'

'Oh, you're full of all the cheery thoughts of the day, Angelique. But here's my question, even allowing for what you just said: if he survived Dubh Ardrain, and somehow got away, maybe with another new identity, why would he break cover *now*, after all these years? And what's he been up to in the meantime?'

'I've asked myself the same thing. The scary answer is "planning".'

They reach the Charing Cross Road and turn right, crossing away from Burleigh Mansions where one Thomas Stearns Eliot used to keep a crash pad. 'Humankind cannot bear too much reality,' he once wrote. Unfortunately, its tolerance for reality TV is proving less fragile. As they pass Borders bookstore, Angelique observes that there's no place in the window for the one-time local boy: the inset bay is taken up with a display promoting Darren McDade's book, *Who's the*

Daddy? He got paid a shudderingly huge advance for his cobbled-together compendium of populist rantings, only for the very public that was believed (at least by one literary editor, after a catastrophically good lunch) to be hanging on his every word, to widely baulk at the prospect of paying to read the same columns they had already coughed up for when they bought their papers. Perhaps this was a measure of his success in dissuading them from the tree-hugging practices of recycling, global warming being a muesli-eaters' myth, apparently. Now, however, he was finally earning those imprudently generous royalties.

A few yards further on, the music section has all three Four Play albums (debut, flop follow-up, greatest hits) prominently stacked close to the doors, and one of their tracks playing on the in-store stereo. Needless to say, it isn't 'You're Dynamite'. It's going to be a bugger of a long time before a radio station, a shop or any publicly accountable organisation ever plays that again. The bookies have stopped taking bets on their 'new' single going straight to number one on next week's chart. The song is called 'Gone But Not Forgotten'. It's a previously uncelebrated number from the less successful second album, a cue to hit the skip button on most fans' CD players, now elevated immeasurably by the sheer marketing potential of its unintentionally poignant title.

'I don't know, Angelique. I'm aware I'm trying to talk myself out of worrying, but the other thing that doesn't quite fit for me is the targets. He was a mercenary—he killed for cash and, let's not forget, because he liked it—but he didn't choose the targets. This psycho we're dealing with now strikes me as being about something very, very heartfelt and personal.'

'There's some classified stuff I can't divulge, Ray, but believe me, Simon killed people for his own heartfelt, personal reasons as well as for professional ones. How do you think he started? He didn't wake up one morning and decide to try his hand at being a hitman.'

'Okay, granted, but allowing for that, what do the targets say about the perpetrator? Because I don't think they say "Simon". I mean, Nick Foster? Crappy teeny-pop bands? It's a bit of a comedown from taking out power stations, cruise liners and military bases, is it not?'

'You told me yourself how ridiculously serious Simon was about music, as well as how angry about the world not quite falling at the feet of his own talents. Don't you have any recollections of him ranting about the Nick Fosters and Four Plays of the time?'

'It was the Eighties, and the Nick Foster of the time was Nick Foster. But no, not really. The kind of music Foster churned out was literally beneath contempt. Saying you hated manufactured bands was like saying you hated wasps. Us music-obsessives had our prejudices and irrational dislikes, but it took 'serious' music to pique serious hatreds. Simon really, really fucking hated The Smiths. If it was Morrissey strapped to that giant amp, you'd have a positive ID on Simon right there. Or maybe the guys from Chambers of Torment, because they truly represent the boat he missed. Back when I knew Simon, Nick Foster wouldn't even have registered on his radar.'

'Foster's been a lot harder to miss in recent

years,' Angelique reminds him. She doesn't want Ray getting entrenched too deeply in his understandable attempts to convince himself that Darcourt has not returned, as she needs his mind to be open when he gets to the post-production house. She's got a more compelling argument in her armoury, but she's saving it for the right moment.

'Not so hard to miss if you don't live in the UK,' Ray counters.

'Don't be so sure, Raymond. Christ, I've been living in Paris for five years: it doesn't guarantee you an escape from hearing about *Big Brother* and *Bedroom* bloody *Popstars*. If anything, being somewhere else makes you even more resentful that it's still able to find you. Can't you see someone as deranged as Simon deciding this was one way of stopping the rot?'

'I can't see Simon tuning in via satellite from some mainland European bolthole just so he can stoke up his rage. And that's one of many reasons why McDade doesn't add up either. Can you picture him getting the *Sun* or the *Daily Mail* delivered to his underground lair or wherever he'd have been hiding? Plus, look what was done to the guy—that was a lefty-liberal revenge fantasy come true. Revenge fantasy might be Simon's style, but the lefty-liberal part . . .'

'Sure,' Angelique agrees. 'I appreciate that if you wanted to plot the human spectrum of social conscience, you'd stick Mahatma Gandhi at one end and Simon Darcourt at the other, with maybe the distance between him and Donald Rumsfeld being how you calibrate the scale, but—'

'No,' Ray interrupts. 'What you don't appreciate

190

is that Simon wouldn't be *on* the scale, because the scale would be irrelevant to Simon. The only political leader you could identify him with would maybe be Margaret Thatcher, simply because she said there was no such thing as society. That's pretty close to Simon's perspective, though it would be more accurate to describe the way he sees it as there's one person who matters and six billion supporting players of varying levels of minor relevance. Simon only reacted to issues when they directly affected him, and even then he never saw it as politics, but as a personal affront. He wouldn't give a fuck what Darren McDade said about penal codes or asylum seekers, so he wouldn't be concerned with driving home any points about either of those issues.'

'Granted, but I think you could be missing the point he *was* driving home. It wasn't about penal codes or asylum seekers, it was about humiliation, and this "lefty-liberal fantasy" as you put it was the means by which he felt McDade would be most humiliated.'

They take a left and they're on Wardour Street. The post-production facility is in sight, only three pubs and four porn stores away. Time for the money shot.

'The lefty-liberal part was incidental,' she continues. 'But the revenge part was genuine. McDade wrote some strong stuff about Darcourt.'

'So did every columnist in the country.'

'True, but only one wrote about Darcourt's father.'

Ray turns his head sharply and slows his stride. Impact confirmed. Angelique follows up the blow.

'I had someone search through all of McDade's

rantings post Dubh Ardrain, and as well as the standard outrage, he uses it as an angle to vent one of his many personal prejudices. "Simon Darcourt was born in Scotland,"' she quotes. ' "But the Jocks don't need to apologise for him. His father was French and a failure, though the latter part goes without saying once you've established the first . . ." Then blah blah, planting trees on the boulevards to let the Germans march in the shade, surrender monkeys, etcetera, etcetera.'

Ray says nothing for a few moments. They stop outside the audio house.

'Are the targets saying "Simon" now?' Angelique asks.

'Let's find out.'

*   *   *

Angelique's mobile vibrates and she glances at the LCD before excusing herself, leaving the suite to take the call while the sound engineer prepares another sample for Ray. She was planning to bail once Ray was ensconced, having been warned that the process could take a while. She just had to deliver him and explain what was required; he wouldn't need a babysitter the whole time. Nonetheless, she had sat there and observed for the best part of an hour, turning down the exit opportunities afforded by two previous calls. However, as both of these had been from her mum, she couldn't say whether she had stayed put out of curiosity or convenience.

Two of the three voicemails left while her phone was off during the press conference had been from her mum too. The first had been an excited call to say she had just seen her on Sky News, the words 'live press conference' failing to register any implications for Angelique's availability to take any calls at that particular moment. The second, left less than ten minutes after the first, was more familiarly accusatory and unsubtly guilt-laying, starting off by asking why she hadn't phoned back regarding the first message, then moving on to slating Angelique's failure to inform her she was back in the UK, before inevitably demanding to know when she'd be up to Glasgow for a visit.

What a great idea, mum. A trip to Glasgow: that would help fill in some of the countless hours she had spare and didn't know what to do with. And what a morale-booster it would be too. She and her parents could gather round the dinner table and have a special celebration to mark the five-thousandth time of her mum sympathetically—ie not remotely optimistically—asking her whether there was 'someone special on the horizon'. Then they could all drink to commiserate with Angelique upon the ongoing loneliness and futility of her life without companionship or the even distant prospect of ever hearing the patter of tiny feet. Angelique had tried just as many times to explain how her job hadn't made it very easy to meet prospective partners, nor was it the most stable base for young family life.

'There will always be crooks, there will always be police and there will always be jobs, Angelique,' her mum would respond. 'But some things *won't* wait forever.'

As if she needed any more pressure in her life, she had been getting her own biological countdown thrust in her face like it was the old Irn

193

Bru clock above Central Station.

It used to piss her off because she thought her mum didn't appreciate what she was involved in and didn't understand what she wanted from life. Now it pissed her off because she realised her mum had always known better than she the precise value of what she was involved in compared to what she really needed from life.

She hadn't told her parents she was coming over to London on this attachment, far less that she was planning to quit the force. Angelique still felt equipped to handle Simon Darcourt one more time, but having to hear her mum say 'I told you so' was a prospect she was nowhere near strong enough for.

The call is from Dale. She's about to tell him it's early days on the voice-work with Ray Ash, but he doesn't even ask.

'Get to a computer,' he says. 'Our boy's back online.'

'Just gimme a minute,' she tells him. 'I'm down in the basement. I need to get upstairs to the main office. More videos?'

'Oh, we're going way upscale now. It's a fucking multi-media extravaganza, and speaking of media we'd better brace ourselves for a force-ten shitstorm. He's unleashed a multi-headed hydra on us.'

'Less girly squealing, more details, please,' she says, walking into the office. 'Sir,' she belatedly adds. She glances at one of the engineers and gesticulates towards a free monitor by way of asking permission. He gives her a nod and she hits the on button. A fan starts to whir and the hard drive lets out a fart as the system boots. Sounds

194

like an old man waking up in the morning.

'He has indirectly made the Black Spirit claim public,' Dale continues, 'while simultaneously setting his cap—or should that be black cowboy hat—at a place on next week's hit parade.'

She thinks of her words to Ray earlier, reminding him how angry Simon had been about the world not quite falling at the feet of his own talents.

'He's released that track, hasn't he?' she says. 'His version of "Hurts So Good".' Low-quality rips of the song lifted from the videos and from recordings of PV1's party coverage were already in circulation on the net. Its sick-joke value, combined with a sense of the forbidden, gave it an iconoclastic kudos that had guaranteed its proliferation. Those properties would be multiplied tenfold in an 'official' cut, released by the artist himself.

'He's called it "Hurts Like Dynamite", and it's credited to "The Black Spirit versus Nick Foster and Four Play". It's a polished-up version of the combination of the two songs that formed the soundtrack to the Nick Foster and Four Play murders. Their vocals are still on there, as are Foster's digitally manipulated, in-key cries of pain, though he's faded down the Four's descent into hysterical screaming and replaced the explosions with cymbal crashes.'

'Going all soft on us?' she says bitterly.

'No. You ask me, he's done it so that it's more listenable. He doesn't just want a few sickos checking it out for thrills—he wants people playing it on their iPods, dancing to it in the fucking clubs. He's simultaneously released it on to several mainstream download services.'

'How can he do that? He doesn't have a record company or a music publisher.'

'You don't need one to license your track for download; not a legitimate one anyway. Just the online semblance of a company and an account for your share of the take to be deposited. Our people have been all over it since the track appeared. The company is a phantom, but the account is the real kicker: the bastard's set it up so that every penny goes to a *bona fide* youth music charity.'

'Does he think that will stop us pulling the plug on the downloads?'

'No, he's covered that part elsewhere. I think this is about him ramming the point home. There's a warped self-righteousness running through everything he's doing. Naturally, the charity will be mortified, but they won't be able to refuse the money as it's going straight into their bank. They can take it out again, of course, but they can't exactly give it back to him, so Christ knows what will happen there, other than a lot of public debate, discussion, publicity; in short, everything this headcase wants.'

'So how has he covered the issue of us—'

'You got web access yet?'

'Just about. Haven't seen anything start so slowly since Rangers sold Mikhailichenko.'

'Okay, launch the browser and key in the following.'

Dale dictates a numerical address. Angelique keys it in and is routed to a black web page displaying the Rank Bajin image at the top above an embedded video window, which immediately begins to buffer its content. Beneath this are three thumbnail images, identified with underlined hyperlink text as Anika, Sally and Wilson. There is also a button marked 'Forum'. She clicks on each image in turn, launching new embedded video tabs, all of which also begin buffering.

'I'm not seeing anything yet,' she says.

'Which should tell you how many people are already drawing on the same bandwidth. Give it a moment.'

Angelique clicks through the tabs, and after a few seconds they begin to successfully stream silent video. She sees the three members of Vogue 2.2: each looking tired, scared, tear-streaked and sweatily dishevelled, each isolated in a small, identical cell, fully in shot. Apart from the human occupants, each cell contains only three items: a thin, pale grey mattress in one corner of the floor, a bucket, placed as far from the mattress as space allows, and, mounted high on one wall out of reach, a tubular canister.

'Are these live streams?'

'Far as we can tell. All except the main page, which is a recording—more overdubbed vocals from our host.'

'It still hasn't loaded.'

'Yeah, it's drawing the most traffic—everyone goes there first before selecting any of the streams.'

'Oh, here it comes.'

The video window displays a short montage of Vogue 2.2: miming onstage as a trio, 'performing' individually on *Bedroom Popstars* and glimpsed in a fast-edit moving scrapbook of their countless tabloid and magazine clippings. Then the killer's now-familiar altered voice comes through the speakers.

'Welcome, all you insatiable, amoral, vicarious thrill-seeking voyeurs. You've come to the right place. Though, can I just offer a big apology to all of you out there who don't fit that description, and who have only logged on out of genuine concern for the plight of my three contestants. *You're* not voyeurs, or amoral vicarious thrill-seekers. I acknowledge that. You're hypocritical lying cunts. Accept what you are or log the fuck off, right now.'

'Christ, how long does this keech go on for?' Angelique asks Dale. 'Can't you just talk me through it, bring me up to speed with a non-self-satisfied gloating prick version?'

'Might as well hear it from the horse's arse, Angelique.'

She sighs, keeps watching the screen. The view cuts between the three cells, showing earlier footage of Anika, Sally and Wilson.

'. . . very special guests, all of whom have proven themselves real champions on another televised popularity contest, so who better to take part in my new, higher-stakes reality game show? A big welcome to Wilson, Anika and Sally, who are, as of now, the first contestants on *Dying to be Famous*!'

'Oh no,' Angelique breathes.

'The rules are simple. Each contestant is locked in an airtight cell with, as you can see, a tank on the wall. In the tank is their air supply—let's call it the Oxygen of Publicity. To keep getting oxygen, they need to get publicity, which should of course be a piece of piss for these three, especially under such terribly moving circumstances. But here's the catch: the one who gets the fewest mentions on TV, who clocks up the fewest column inches dedicated to him or herself over each twenty-four hour period, gets the least oxygen. There'll be enough so that they can each recover from a bad day and maybe rally the next, but three, maybe even two days at the bottom of the vacuous-gossip chart, and it's off to the big talent show in the sky. But don't worry, it won't all be in the arbitrary hands of those self-important wankers in the media: this is a British popularity show, after all, so *you* can get involved too. Go to the forum below, and give us *your* opinion on who should die first. But remember to give a reason, otherwise your vote won't count. Any reason you like—don't worry if it's just blind and irrational prejudice—that's good enough!'

The screen fades to black, then the Rank Bajin cartoon figure appears, approvingly holding one thumb up to camera. A triangular play icon appears in the centre, offering the option to view the video again. Angelique wants to hit it with a hammer.

'I am assuming the only reason this fucking abomination is still live is because you're running a trace.'

'Yes and no. We received a message, via email, telling us . . .'

'Same deal as the Tivoli. I get it. If we don't play the game, he takes his ball and goes home.'

'That's it. If the site or the feeds get pulled, he kills the hostages; same deal with the music downloads.'

'He's going to kill them anyway.'

'And he knows we know that, but he also knows we can't act on it. We have to let him have his fun.'

'What about the trace?'

'The message claims that he has monitoring in place to detect any attempt to trace the source location. If his detection measures pick up any trace activity, he kills one of the hostages.'

'He's bluffing. Is that kind of detection possible?'

'Our geeks are on it. They say just from a cursory look at the set-up that he knows what he's doing: tentative initial traces show the signal coming simultaneously from a hundred nodes on five continents. Christ, even the email he sent was carrying the digital equivalent of a postmark from half the major cities on the globe, and appeared to have originated—get this—from Cromlarig, in the north of Scotland.'

'Nearest town to Dubh Ardrain,' Angelique acknowledges. 'Very fucking funny. So if he can spoof trace-route signatures, can he do what he's claiming?'

'Detecting our attempts? Geeks say not likely. Not theoretically impossible, but chances are he's bluffing to deter us from trying to sniff him out.'

'Would he chance being traced just to keep his show live? He could as easily just upload more videos on a regular basis. Less risky.'

'Frankly, I don't have a clue. I'm just going with what the geeks—' There sounds a double-beat tone on the line. 'Sorry, have to stick you on hold.'

Angelique is left staring at the screen, just a regular electronic pulse on her mobile to reassure her that it is still connected to something. As the seconds turn into minutes, she can't help but click on the open tabs, and finds herself gazing at the three prisoners. Perhaps it's the camera angle, but they all look somehow really small. They also look bedraggled, disorientated and exhausted, indicating they have barely slept in days. The sense of voyeurism is curiously enhanced by the absence of sound, making it seem like security footage of people oblivious of the watching cameras, though the three of them seem not so much oblivious as ignoring. The occasional glance indicates that they know about the cameras—so they're not hidden—but they seem indifferent towards them, like they offer intrusion but not communication. That's when it occurs to Angelique to wonder about the lack of sound. Is Darcourt scared they could communicate something about their location or about their captor? But then she gets it: another of the bastard's sick jokes. They got where they are through miming. They never wanted a voice then, so he's not giving them one now.

Dale's call resumes on her mobile. Dale sounds younger than he looks, a certain energy and enthusiasm in his voice that shaves a few years off, particularly in a profession with a tendency to advance world-weariness in anybody's register. Of course, it's always possible his voice is bang-on for his age and it's the strains of the job that have made him look ten years older. Even if so, that extra ten still only makes him look early forties, and good for it. But let's not even begin to go there . . .

'Sorry about that. That was the head geek, looking for my green light to commence some hopefully very gentle probing. They're running whatever it is right now, proceeding on a steady-as-she-goes basis. If they encounter anything they don't like the look of, they'll hold off. But as you say, he's going to kill them anyway. It's a chance we

201

have to take. It's not like we can . . . fuck. Fuck.'

'What?'

'How is it your end?' he asks hurriedly, anxious. 'Have you lost it as well?'

'Lost what?' Angelique asks, then notices that the video window has gone blank. She clicks through the other tabs: they're all dead; hits Refresh on the main page, just gets a 'connecting to . . .' message.

'Feeds have all dropped,' she reports. 'Index page is getting ready to 404 me any second now.'

'Jesus Christ. I'm calling the geeks again. Keep refreshing. And hit Shift-F5 to clear the cache and fully reload.'

'Sir,' she affirms.

The on-hold beep becomes like an electronic echo of her own pulse as she hits Shift-F5, Shift-F5, except her own pulse does not stay at such a digitally calibrated steady oscillation. She thinks of the three scared faces she has just been looking at, images so vividly ordinary they could each be in a room next door.

Shift-F5. Connecting to . . .

Shift-F5. Connecting to . . .

Then all of a sudden her whole body jolts in her chair as the index page reappears: the Rank Bajin logo where it was at the top, the main video window below, and underneath that, the explanation for the downtime: he's added three mirrors underneath each of the hyperlinked names, adding more bandwidth to cope with what he anticipates will be an ever-increasing load.

She issues a steam-vent sigh, almost laughs at the relief, but it's relative. The situation remains merely dire, as opposed to fatally irretrievable.

'I'll be doing well to get through this without a heart attack,' Dale says. 'Though at least an MI would save me from issuing a new statement on all of this. The news channels are about to go insane. I'd like to hold off for a few hours, but I've got to get the message out ASAP to any would-be hackers not to attempt any private investigations. It would look a sight better if we had something to offer by way of suggesting we have a lead, other than merely a corroboration that this guy really is Darcourt.'

'I'll go downstairs and see how we're doing on the voice.'

'I expect I'll be making a statement within the hour, but if you get something, call me, even if I'm on-air at the time.'

'You got it, sir. Oh, but one thing before you go: corroboration? Do you mean the Cromlarig link in the email? Because that's even more public domain than the Rank Bajin picture.'

'Shit, can't believe I forgot to say. I meant the email, yes, but not the Cromlarig part. There was a hotlinked image at the bottom of the text. Photo of Halle Berry.'

'Christ. From her best-known role?'

'His way of saying he got our message.'

Angelique knew what the image on the email would show without having to see it: a dark-skinned female in a skin-tight black bodysuit, a similar sight being one of the last things Darcourt and his men clapped eyes upon before being taken down at Dubh Ardrain.

Halle Berry: most famous as Ororo Munroe, aka Storm.

The X Woman.

203

Angelique walks near silently down the stairs.
She's not making an effort to be stealthy; it's just
the way the place has been designed to absorb and
muffle sound. The carpet on the stairs is hard-
wearing and feels stiff beneath her shoes, but it
cushions every footfall, and the walls either side
hug the narrow passage so tightly as to prevent
much reverberation. As she turns into the corridor
at the bottom, she can see Ray through a double-
glazed window. He looks precipitately drained, like
he'd have been fine if she came down here twenty
seconds ago, but now the colour has just been
flushed out of him as if somebody pulled a plug.
The description that would leap to most observers'
minds is that he looks like he's just seen a ghost.
Angelique is better informed: she knows it's that
he's just heard one.

She waves to Gary, the engineer, requesting
permission to enter. He gives her a nod and she
reaches for the door. As soon as she pushes the
handle and breaks the rubber seal around the
frame, she can hear the voice all around. Ray turns
to look at her, his face a mixture of fear,
uncertainty and not a little accusation: what have
you done to me? He looks like he might be sick.
Angelique doesn't have time to offer assurances, as
within moments she's busy dealing with her own
reactions to the voice, now uncloaked of its digital
camouflage and stalking her memory like a
predator.

Her hand goes to her chest, because that's
where she suddenly feels it, all over again, like

somebody just opened the vault where her subconscious had locked up what it couldn't deal with in that terrible moment. Physically, she's standing in a sound-editing suite in Soho, but her mind is back at Dubh Ardrain, lying on the floor in agony and shock. The Kevlar stopped the shotgun pellets from penetrating her skin, but the blow itself had been point-blank and it was her ribs that absorbed the force. She never actually lost consciousness, but felt swamped, overwhelmed by brutal sensations: pain, disorientation and a deafening ringing in her ears from the muzzle-blast. She remembers his voice—*this* voice, this voice she can hear now in Soho—becoming audible and intelligible as those sensations gradually dissipated. She cannot hear it without experiencing a sense-memory of that pain and fear, but buried a layer beneath there is also a sense-memory of what came immediately after: of hate, of anger, of burning resolve. Here in the editing suite, she feels the bile rising, her heartbeat that bit more insistent.

'I'm going to have to hurry you, Nick. I realise the lyrics are utterly vacuous and probably indistinguishable from a thousand other shitty songs you wrote, but *you did* write them, so why not take a stab.'

There it is: that narcissistic conceit and self-satisfaction in the voice, that detestably smug arrogance. As they used to say when Angelique was wee, if he was made of chocolate he'd eat himself. All but chuckling at his own cleverness.

He had stood there with Angelique flat on her back and Ray on his knees, the situation back under his control, the whip hand his again, yet the fucker still felt the need to sing when he was winning:

'And what have *you* got to wank about, in your ordinary, anonymous little life? Tell me that. What the hell have you achieved? A fucking school teacher. Wife, mortgage, and a kid now, I understand. You really shine out in the crowd, Ray.'

He hadn't changed. Same sadism, same gloating, same complacency, same weaknesses. And when she got hold of the fucker, same result.

This time, though, she'd make sure there was no escape. She was going to slap the cuffs on him personally, and she was going to make sure he was in no state to resist, far less disappear again. Before she clapped him in irons, she was going to see him on *his* knees. See him humbled. See him defeated.

See him bleed.

*   *   *

Angelique calls Dale. In the background she can hear a hubbub of chatter and the scrape of large objects being hauled around a polished floor. The Tivoli. She hopes she's not too late.

'You're a life-saver,' he tells her, though she suspects they both know that this is a purely metaphorical compliment. The influence of the new voice recording on the fortunes of Anika, Sally and Wilson is unlikely to be so decisive, and she says as much.

'We have to take encouragement where we can find it in a shitstorm like this,' Dale responds. 'This at least lets us change the agenda for the next news cycle. The latest angle becomes positive progress about our new lead, us getting somewhere rather than looking literally clueless against a background of hysterical reaction to this latest horror show. Like it or not, this gig is going to be as much about managing the news as about investigating the leads. Playing the media is not a strategy, it's a necessity, because the fucker we're up against has made the game so, and we have to make sure we play it better than him.'

'I know,' she concedes. 'It's just a hell of a culture shock after several years of covert operations.'

'Never knew when you had it good, huh?'

She gives a short, wry laugh. 'I'm firing over an initial sample of Darcourt's voice for immediate broadcast,' she tells him. 'Ray and Gary, the sound engineer, are fine-tuning the settings just now so that they can issue a complete set of all the recordings, but a rough-cut preview should suffice for the press conference.'

'Good work. When they're all done, we'll get them online for download, get the link on to the news, *Crimewatch* and hopefully the front page of the BBC website.'

'Speaking of websites, any news on tracing those live streams?'

'No, and don't be checking regularly for updates, because I'm informed it will be slow going. They've encountered firewall after firewall; not so much a needle in a haystack as like trying to find a particular strand of hay when each wisp leads to another entire haystack.'

'There must be something they can do: Darcourt's smart but he's not a computer genius.

Surely their command of the technology ultimately has to be better than his.'

'There *is* something they can do, but it's what the alpha geek called BFI. That's hacker jargon, stands for Brute Force and Ignorance, meaning in programming terms a solution that gets the job done, but in an unsubtle, messy and haphazard fashion. The problem is, we need them to be stealthy, so BFI is not an option. They're working on alternatives, but it's not going to be swift or elegant. "Like kicking dead whales down the beach," was the phrase he used.'

'Sounds almost as much fun as manning the phone lines when we get the public response to the voice tapes. Commiserations to the officers on nutter-fielding duty.'

'Agreed, but where this could prove more useful is that it might make it harder for him to operate. If we can get the public familiar with his real voice, then they'll recognise it if they subsequently hear it for real. It's not going to stop him from being able to grunt a few words to the cashier when he's paying for his petrol, but if he wants twenty Regal King Size or a new order of oxygen canisters . . .'

'Then he just has to put on a different accent.'

'Shit. True enough,' he concedes. 'But on the bright side, all it would take is one person to genuinely recognise that voice and the whole game could change.'

'A big if, though. The problem is, how many people will have heard Darcourt's voice recently enough to recognise it? This is not someone who makes a lot of new friends. No matter how solid a new identity he's created for himself since Dubh Ardrain, he'll have been laying very low. We're not

208

going to end up interviewing some next-door neighbours who'll say how normal he seemed, very polite, wouldn't have suspected a thing, always made sure his rubbish was out on time. More the extreme end of the "kept himself to himself" part of the public-profile spectrum. I can't think there's many people left alive who'll have heard his voice since his *first* supposed death, over a decade ago.'

'Like I said, Angelique, it might only take one to change everything.'

THE GOOD SHIP BLACK & DECKER

'All of the British-based news channels are running with the clip on their headline cycles, sir. Sky and BBC News 24 are offering playback on demand via their digital services. Complete, unexpurgated versions of all the audio tracks are expected to begin appearing on the BBC's website within the next two hours.'

'An exciting twist to an already intriguing story, wouldn't you agree?'

'As I reported to you previously, sir, our sources close to the investigation indicate the police have had reason to believe it was Darcourt since the first murder, the English journalist. That is why they brought in this woman, this Angelique de Xavia. She had been working out of Paris—'

'Under that nuisance Dougnac, yes, you said.'

'And if I may remind you, regardless of the official reasons, she was drafted into the investigation because she has first-hand experience of Darcourt. She was instrumental in the Scottish fiasco of 2001.'

'Quite. An extremely expensive mess. Unavoidable from time to time, but fiascos such as that serve to remind us why we insist on fifty per cent up front and the rest payable upon completion.'

'Indeed, sir.'

'The only consolation was the absence of wind-borne contamination. There was, of course, the not inconsiderable cost of having Mopoza assassinated before his verbal incontinence and mental instability combined to leak something potentially toxic, but ultimately, we didn't sustain any structural damage, only financial. The latter is always more easily repaired. But that's what concerns me about Darcourt's reappearance. There is a histrionically self-righteous agenda at play here, no code of professionalism. It has disturbing implications for his, shall we say, moral neutrality. One cannot help but be reminded of the Lombardy incident.'

'I concur, sir. In retrospect, there is a certain consistency of style and, more disturbingly, motive. Not the specific motive itself, but the very fact of there clearly *being* one. If Lombardy, indeed, was Darcourt, then its overtones of vigilantism dictate an express requirement for decisive action.'

'The Lombardy incident was how many years ago?'

'Four years, sir.'

'Darcourt survives the disaster in Scotland, then sensibly stays buried for all these years but for—if in fact Lombardy was him—one piece of work. Indeed, Lombardy would only emphasise that which perplexes me, because he left no clue or claim to its authorship, and vanished once more. Four years after that, more than six years after the Scottish calamity, all of it spent in perfect anonymity, why suddenly break cover, and in such an ostentatiously, *spectacularly* public way?'

'I share your perplexity, sir. He could not have done any more to bring attention to himself, yet it is difficult to see what motive would lie behind such a sacrifice of exposure. The risk-benefit equation simply does not compute, which suggests an erratic aspect that is dangerously, dangerously volatile. But be assured, we already have operatives in place . . .'

'I am assured of that, Bernard. I *assume* that kind of assurance, otherwise I would employ someone else. My disquiet remains, however. Something here does not add up, and thank you for pinpointing it: the risk-benefit equation. What is Darcourt about? What has brought him so careeringly into the open after all this time? What could be so valuable in order that he would expose himself with such apparent recklessness, that he would so heedlessly risk . . . Unless . . . Oh my. Oh my.'

'Sir?'

'A moment, Bernard. Hmm. Unless my instincts are wrong—and they seldom are—then this could be entirely as bad as we might imagine. It is imperative that we intercept Darcourt.'

'Yes, sir. I will inform the operatives that the mission is now live.'

'Standard operatives will not suffice. I believe Simon has a secret: a secret that makes him the single biggest danger on our radar. The potential for damage cannot be underestimated. We will need to engage a guarantor . . .'

FIRST CONTACT

It's day four in the Black Spirit house, and things aren't looking good for Anika. She's slumped on the floor with her back to the wall, eyes closed, her matted hair stuck to her cheeks, sweat-soaked clothes clinging to her frail frame. She hasn't moved in close to an hour. Her breathing is so shallow, it's hard to tell whether she's doing it at all; only the beads of fresh sweat on her face indicate that she is still alive.

Angelique feels a swelling in her throat. She can't afford this, can't let herself feel this, not now. It's too soon, and if she lets it happen once, she'll be fighting against it the whole time. She needs the distance, the detachment. Needs a couple of drinks and some sick jokes to put it all at a remove. Instead, it's inescapable. There's monitors everywhere, the three largest of them suspended from one wall, each permanently dedicated to one of the live streams.

'Can't we turn those fucking things off?' she asks rhetorically, spinning on a heel to at least face away from the triptych for a moment. She finds herself facing Dale, who is standing beside alpha geek Julian Meilis, watching his monitor over his shoulder. Dale looks up in response to her outburst, gives her a sympathetic grimace. He understands, he feels it too; so much so that his solidarity makes her feel self-indulgent. It's like a siege, and nobody can afford the luxury of histrionics.

They're in the Operations Centre; no mere

213

Situation Room on a case like this. It's a place she's been doing her best to avoid, due to the widescreen views of the ongoing snuff feature that dominate the centre of the room, three open grates upon a sewer of the mind. Right now, however, attendance is mandatory, as they watch and wait and the geeks ready themselves, finally, to act.

Anika has finished bottom of Darcourt's publicity tally for two days in a row and has therefore received the least oxygen. Nobody is sure how he is regulating it, but it is believed that the contestants' allowance is released gradually, so that the levels remain consistent, with Anika's consistently lower—then lower again—than her erstwhile colleagues, now competitors.

He has also allowed them each a bottle of water per day, and two Mars bars. Someone suggested that the choice of food was supposed to maintain their energy levels, but Dale was keen enough to observe that Mars was the official sponsor of *Bedroom Popstars*. 'They should all be grateful it wasn't Ryvita,' he remarked. Similar speculation focused upon the consistent choice of Evian as their sole source of hydration, and the product's current, previous and even tangential associations were analysed for significance until Angelique pointed out that the name spelt 'naive' backwards.

Wilson bottomed out on the publicity score the first day, but has subsequently rallied, leaving Anika in her current and worsening plight. The op-ed commentators are saying it's because she's not from an articulate and photogenic middle-class family, and thus her friends and relatives are less adept at procuring those now life-giving column inches. However, said commentators are themselves redressing the balance by generating this copy, much as they did after day one, when they opined that Wilson had found himself bottom of the heap because he was black. Commentary and posturing around this issue certainly helped rally Wilson's fortunes, to the extent that the message-board Darcourt set up is now full of comments from white van man, decrying Wilson's improved fortunes as evidence of positive discrimination within the politically correct and liberal-dominated media.

For all that, Wilson still lags behind Sally, the consistent leader in the publicity stakes and thus the favourite to survive the longest. Numerous theses have been postulated to explain Sally's favoured status, but the message-board consensus is apparently that it's because 'she's not wearing a bra so you can see more of her tits'.

Yes, Darcourt is getting his dark circus, his 24/7 live gameshow in which *everybody* is participating, whether they intend to, whether they even realise it, or not. Every piece of watercooler gossip is a contributory factor, the beat of butterfly wings that can affect the prevailing wind and ultimately determine how much of the air finds its way into each hermetically sealed cell.

But of course, as well as butterfly wings, there are also giant turbines, for nothing is so capacious in generating hot air as the British tabloids. They can even blow off about their self-righteous efforts *not* to blow off. One of them claims to have been doing an official Coverage Count before going to press each day, and amending copy so that all three of the contestants get an equal tally of column inches. 'We're not playing the Black Spirit's sick game,' they announced, while amending their own content as dictated by the rules of the Black Spirit's sick game, and covering said sick game on more than a dozen pages. One could at least admire this particular paper's moral fortitude in attempting to avoid influencing Darcourt's lethal gauge, and only a cynic could suggest that this self-righteous posture was a sour-grapes response to three rival titles having signed up each of the victims' families on exclusive deals.

The contestants' own efforts to tip the scales in their favour have been understandably limited; in fact, their lack of means to communicate has already caused Darcourt to liven matters up with the very innovation Meilis and his fellow geeks are waiting for as their cue. The three inmates, bereft of voice and denied anything with which to write, have all independently (though on varying timescales; Sally a full twenty-eight hours behind Anika, herself half a day behind Wilson) arrived at the realisation that someone watching out there must be able to lip-read. Experts have subsequently been brought in to interpret from recorded footage, but yielded nothing constructive in terms of assisting the police search, nor individually compelling from a publicity point of view. Expressions of fear, growing desperation, disorientation and various messages of love to family members have been the extent of their visemally conveyed contributions. The irony of their having effectively nominated themselves for this horror via their lip-synching abilities has remained unremarked upon in the media, presumably on grounds of taste and sensitivity rather than it failing to strike anybody, as it is clearly a point Darcourt intended to convey.

Angelique feels her phone vibrate. She takes it from her pocket with the sort of reflexive swiftness she normally reserves for blocking an incoming blow, and holds it up like she's flashing her ID. It is, after all, a form of pass: her ticket out of the Operations Centre for a moment or two.

'Hello,' she says, to underline that it's an incoming call.

'Ah, ha ha!' cackles a voice with a sense of triumph as amused as it is mischievous. 'Got you at last.'

Oh shite. It's her mum.

'I'm out with Arlene and I thought I'd give you a try on her mobile,' Mum explains, still giggling. Angelique winces to herself. Her mum's unspoken implication is that she took the call because she *didn't* know who it was from, and she's right. Mum shouldn't take it personally; or at least not quite as personally. It wasn't that she'd rather take an unknown call than one from her mother; not always anyway. It was a matter of cop life, the opposite of a social one. Normal people take calls from the numbers they do recognise and give a reduced priority to ones they don't. Angelique, however, could never ignore that unknown number in case it turned out to be the source of her next inquiry: the anonymous tip, the ransom demand, the breakthrough lead. Never mind the unwary public: the mobile-phone scam artists would really clean up if they could get a list of coppers' numbers. They'll take any call, ring back any number, especially in a situation like this one.

What's even more annoying than her mum's

successful subterfuge is that right now is the one time Angelique *would* have happily—gratefully—taken her call: any excuse to get away from those monitors.

'I won't keep you, darling, I realise how busy things must be—not to mention how horrible—but I had to try and get in touch before we go.'

'I'm sorry, Mum, as you can maybe appreciate, it's not been easy to . . . hang on, go? Go where?'

'That's what I've been trying to get in touch about. A round-the-world trip, for three months.'

'You and Mrs McDougall?'

'No, not Arlene, me and your dad. We're leaving tomorrow, first leg is to Egypt. I'm out with Arlene shopping for some last-minute holiday clothes.'

'Tomorrow? For three months? When did you book this?' Angelique asks, hoping not to register any disapproval or indeed envy in her voice. Three months of escape from the everyday insanity of her world sounds damn good right now.

'We didn't book it,' her mum says with a chuckle. 'We won it.'

'Wow. How? A competition?'

'A raffle. One of those charity prize draws, you know, somebody's always coming round the doors selling tickets for Cancer Research or whatever. Probably shelled out as much in tickets over the years, but finally our number came up. Well, strictly speaking we didn't win it. Turned out the poor chap who did had a heart attack and died a few days ago, no family either, poor soul. We didn't even still have the tickets but the charity had the name and address and it turned out we were second reserve. First reserve couldn't go at such short notice, but well, we're both retired, you and

James long since flown the nest, and . . .'

Angelique's in-built sense of suspicion begins to tingle. She recalls a scam from a decade or so back, whereby the con artists were phoning people up and telling them they'd won a cruise. It would be all-expenses except they had to front a hundred quid for insurance. Once that was banked, every so often they'd get back in touch saying some new technicality required another small payment. Smart psychology: the marks kept shelling out because they had already made several non-refundable payments and, hey, there was a cruise at the end of it. By the time it all added up, they had paid roughly what the cruise would have cost down at Thomas Cook, but the real sting came a week before departure, when the company suddenly ceased trading and all the phone numbers vaporised.

'Wait a sec, Mum. Somebody just phoned you up with this? Have they asked you for any money?'

'No, no. Switch off your paranoia for a moment, Angel, save it for work. It's all legit. The documents are on our mantelpiece, including business-class flights the whole way.'

'Well, congratulations.'

'Thank you. We won't be seeing you for a while, but that won't exactly be a change, will it?'

Yeah. Had to get that in, didn't you, Mum.

'I'm overdue some time off,' Angelique says. 'Maybe I'll fly out and meet you somewhere when this business is all over.'

'That would be lovely,' Mum says, and to her credit she doesn't make it sound like, 'That will be right.' They both know it's not going to happen.

Angelique feels relieved, though: with her

219

parents out of the picture, it's one less thing to feel a guilty failure about. Maybe by the time she sees them again, she'll have that elusive horizon-man in her life and the future worked out. Mind you, it was only round the world, so perhaps not. If it was all the way around Jupiter, brief stop on Io, that would maybe give her enough time.

Angelique checks her watch as she re-enters the room: still ten minutes to go.

She stands behind Dale and Meilis, the latter's widescreen monitor displaying a kaleidoscope of images and information: numbers, read-outs and scrolling code as well as reduced versions of Darcourt's sado-cast.

'FPS is starting to drop off faster,' Meilis states. 'Starting to see the beginnings of some image stutter. Signal integrity is declining across all mirrors.'

The traffic on Darcourt's video feeds has been growing steadily over the past hour because of a growing rumour that Anika is in fact dead. The story started on the *Dying to be Famous* message-board, backed up by a claim that the authorities have somehow engineered a half-hour delay on the transmission because they already know there's been a fatality. Dale has taken a call from Sky News inquiring about this and has said 'No comment' rather than deny anything, with an off-the-record nod to the reporter that she should broadcast his response as 'the police have refused to deny'. This is because the police planted the rumour. It's not true, but the surge in network traffic will assist what Meilis and his team are about to attempt.

Up on the suspended monitors, the three

involuntary competitors continue to suffer in silence. There are only two audio enhancements to their dumb-show. One is when Darcourt announces the day's final publicity league table, which he heralds by playing that fucking Jean Michel Jarre instrumental, which Angelique now has looping in her brain; and the other is what everyone in the Operations Centre is waiting for right now: Mime Time.

After the shock and curiosity effect of the first day, the traffic on Darcourt's voyeur peepholes dropped off, with subsequent surges coming in response to chatter across the various media. Thus instead of the media reacting to people's interest in the show, it's increasingly become the other way around. The problem has been that there isn't much to see: just three people suffering in confined spaces. They can't interact: can't argue, can't form allegiances, can't betray, can't flirt, can't lust. All they can do is sit around and try not to use too much air. Simon isn't interested in putting on a show because the real entertainment, for him, is the media's response. Hence, it's reaction to the story in the media that fuels interest in looking at the 'contestants'. They are supplementary now, figureheads for a thousand surrogate battles being fought on their behalf, though only one of those battles stands a chance of saving any of them, and that's the one the geeks are waging.

However, this means it's a burst tyre for Simon if his wee project gets bumped down the news agenda. He has timed the daily announcement of the publicity tally so that it is fresh for *News at Ten* and its late-evening competitors, but with no developments to report throughout the rest of the day and just the same inert views to look at, the situation has been overtaken as the lead item on the twenty-four-hour channels. Darcourt therefore decided to play a new card, and he knew precisely when to play it, too. Early on day three, he announced that there would be something special worth logging on for at six o'clock, when people are getting in from the day's graft and have access to internet content otherwise diligently blocked in the workplace of every respectable employer. Thus at teatime yesterday, there were more hits on Simon's servers since at any point since the beginning.

The upside for Meilis and his colleagues is that if it was repeated today, the resultant surge in traffic would give them plausible cover to effect a crash of all the mirrors due to apparent overload. This would give them a window to work some kind of baffling techno-voodoo under Darcourt's radar and—with favourable fortune and a strong following wind—hopefully allow them to pinpoint a geographic location for the source of the streams. (In this case, 'pinpoint', Meilis stressed, was merely a figure of speech: 'elephant's-foot-point' being more accurate but less established in the lexicon.)

Hence the rumour leaked to the message-board, hence Dale's 'refusal to deny'. Traffic was already building, which would make it less suspicious when Meilis and his geek squad began knocking over the servers.

Simon doesn't keep his audience waiting; it doesn't do when the next test of their attention span is a mere mouse-click away. Bang on six o'clock, a sound feed accompanies the videos, and all three contestants look up in reaction to the music that has begun playing at their end as well as ours. This is the first time in more than an hour that Anika's eyes have opened. She looks around, opens her mouth, suddenly daring—or able—to breathe deeper. The same thing happened twenty-four hours ago: Darcourt has released a brief oxygen boost in order to pep up all three of them ahead of what's coming.

The music is at this stage merely a thudding bass-drum beat: as yesterday, it's Darcourt's knowingly Eighties-style twelve-inch mix of 'Hurts Like Dynamite', the extended intro allowing him to set the scene and the contestants to prepare themselves as best they are physically able.

'Hello again, all you insatiable vicarious thrill-seekers and hypocritical lying cunts, and welcome to another edition of Mime Time.'

The voiceover is a recording: Darcourt has persisted with redubbing his digitally disguised speech. Angelique had harboured half a hope that he'd save himself the bother now that the truer version was in the public domain, but quickly realised it was still well worth his effort. If the manipulated speech remains the most familiar, remains the voice people associate with Darcourt, then it dilutes the public's memory and awareness of the real thing. Dilutes it to almost homeopathic levels, in fact, given how frequently people are hearing the disguised version compared to the samples Ray and Gary laboriously collaborated to produce.

'Once again, our three contestants get the chance to earn a Breathing Bonus by demonstrating the talent that booked them their place on *Dying to be Famous*. In just a few bars, the music is going to kick in, and the one who mimes it best—decided by *your* votes on the website—will get that special bonus. Yes, they're all competing for a prize of nothing but thin air. Just like being back on *Bedroom Popstars*! And making it interesting tonight is Anika, lagging behind on the O2. I've given all three of them a little extra air so that they can do some serious vogueing. The question is, though, especially in Anika's case: take or gamble? Does she save her breath and hope tonight's tally offers salvation? Or does she give it her all in order to secure that Breathing Bonus and get everybody talking—not to mention writing—about her tomorrow? Let's find out right now on Mime Time!'

Darcourt didn't need to draw any extra attention to Anika, but Meilis is happy he did. The drop in frames-per-second on her feed indicates there's more people viewing her than the other two combined.

'I've not seen lag like this since the dial-up days,' Meilis remarks.

On the suspended monitor, Angelique sees Anika get to her feet. All three are trying to strike poses, looking ready to dance but holding off on any movement until the first lyrics commence. Angelique has seen many, many more brutalities and indignities visited upon people, but in a way this is the most sadistic, the most humiliating.

The song proper begins. Anika tries to dance, though she looks like it's taking all she's got just to remain upright.

Angelique looks at Meilis. 'How much longer do we have to watch this?' she asks.

He holds up one hand—five?—while manipulating the mouse with the other. She wonders does he mean minutes or seconds, then gets her answer as the monitor showing Anika goes black.

'Now watch the dominoes fall,' Meilis says.

The mirror sites for the Anika stream crash in quick succession, blanking off on other monitors around the Operations Centre, as people all over the net attempt to replace the lost feed. Then, once all three sources are down, the effect repeats on first Sally's and finally Wilson's stream as the 'vicarious thrill-seekers and hypocritical lying cunts' connect to the remaining feeds.

There is an unnerving, tense silence throughout this busy room that seems all the more pronounced for its suddenness. Dale looks anxiously at Meilis, his unspoken thought the same as everyone else's: Please tell me Darcourt didn't find that suspicious.

'Don't worry,' Meilis says. 'It worked even better than I hoped. I only had to artificially surge the first server, then network traffic did the rest. I thought we'd need to do it on each of the other two sources to get the ball rolling, but there were so many people suddenly trying to access the same feeds that the servers all crashed by themselves. From Darcourt's point of view, it'll just look like he's a victim of his own success. Now he'll just reset the systems and restore everything, but metaphorically speaking, he has to bend over to put the plug back in, and that's when we can get a free boot at his arse.'

*   *   *

225

Day five in the Black Spirit house. It's seven am and the housemates are all asleep; they're also all alive, which is the good news. Anika won the Mime Time vote close to unanimously, with all but a handful of scumbags logging on to the message-board with the express intention of getting the girl some more air. She has since trumped it by coming top of the publicity tally, her plight eliciting sufficient discussion yesterday to knock Sally into second place.

Meilis and his team don't appear to have moved while Angelique was grabbing a few hours' kip in the serviced apartment she's renting, having been in what the alpha geek calls 'deep-hack mode'. The bad news is that these Herculean efforts have so far failed to provide the location of Darcourt's arse for the promised 'free boot', but Meilis believes they're closing in. On each of their monitors is a patchwork of overlapping windows: scrolling lists of IP numbers, telephone exchange locations, postal addresses, maps, topological diagrams and code, code, code.

'The problem is how little bandwidth is required at source to broadcast these three images,' Meilis explains in between grateful mouthfuls of the Starbucks coffee Angelique brought with her. 'It's further up the stream that it gets mass-duplicated. We've been in a holding pattern for a couple of hours, but we're circling a little closer on every pass.'

'How big is the elephant's footprint?'

'Down to a square mile, but it's a square mile of light-industrial estates near Walsall.'

'Figures. An hour from London, and just off the motorway.'

226

Just then, one of the other geeks' mobiles rings, and he answers it instantly, in contrast to previous habits Angelique witnessed. While in this state of deep-hack mode, she earlier watched the same guy remain focused on his screen, his hands never leaving the keyboard, while the same phone scuttled manically around his desk, vibrating along the Formica to the tune of its *Star Trek* ringtone. She almost answered it herself just to silence the thing, before the caller finally gave up. Now, however, the guy is talking occasionally into his Bluetooth headset—though mainly listening—while still jostling the mouse and cross-referring whatever is on his screen.

'Okay. Okay. Okay. Yeah. Got it. Okay.'

He pats the headset, gently pressing the disconnect button, then raises a hand to wave at Meilis.

'My BT guy in Walsall just came through,' he reports. 'We've got a match, and the best part is, it's a local spur supplying only five premises with active ADSL connections.'

'Brilliant, Rog,' Meilis says. 'Somebody get it on the map.'

'Already there, Jules,' another geek chimes in: Adnan, if she caught his name right. 'It's in the Hillbank estate, north-eastern quadrant. Four places on Mowatt Avenue and one on Beckett Road.'

Angelique glances at Meilis and he gives her the slightest nod. This is happening. She reaches for her phone, getting ready to call Dale.

'What are these places?' Meilis asks. 'Get us some names.'

'Just coming in now,' Rog tells him. 'Okay, we've

got Cook & Co Lithographics, Jacobs Pool Table and Gaming Machines, McArthur Blinds, and a Parcelink dispatch depot: they're all on Mowatt. The one on Beckett is . . . Hodges Bros Catering Supplies.'

'Show me the map.'

Meilis leans across and looks at Adnan's monitor, which he has rotated to give him a better view. Meilis points, but Angelique's angle doesn't allow her to see where. 'That one there. Which one is that?'

There is a brief pause while Rog tabs through several different lists.

'Parcelink,' he informs them.

'That's it,' Meilis declares.

'Parcelink?' Angelique feels compelled to enquire. They're a major firm, practically a household name, and therefore an impossibility in terms of being a front company for someone like Darcourt to secure premises. 'I know they're slow, but do you not think even they'd notice if Darcourt was squatting in their warehouse?'

Meilis smiles by way of acknowledging her query, but instead of an answer, he merely issues further instructions to Adnan.

'Get them on the phone: that depot, not national enquiries. Make sure you're talking to someone on-site, then ask them two things. One, do they have a wireless network. Two, if they stick their head out of the window, can they see an unlet and apparently empty building adjacent to their location.'

Meilis turns to Angelique while Adnan calls. 'In answer to your question, yes they'd notice if Darcourt was squatting in their warehouse. But

228

they wouldn't notice if he hacked their wireless network and piggybacked on to their connection.'

Adnan completes his call.

'Yes to both,' he relays.

'Yes plus yes equals go,' Meilis concludes. 'Time to take that free boot.'

'I'll just look out my steel toecaps.'

*   *   *

Angelique is standing next to Dale, a dozen more armed police taking up their appointed holding positions around the perimeter. She checks her watch. It's almost noon. She feels disconcertingly exposed to be standing outdoors in the April sunshine holding a Walther and kitted in the full body-armour-and-tech-headset regalia. They've checked the sightlines, though, like a conjuror in a new theatre. There's folk working in buildings fifty yards away who will have no idea this happened until they see it on the news.

'Here's where we get to be the good guys for a change,' she says to Dale, checking her firearm.

'What, you saying the general public haven't been unwavering in their appreciation for the protection we provide?'

'Just a little. And the worst part is, often not without reason.'

'I hear you. But they'll be loving us for this.'

'Hmm.' Angelique makes it sound like a murmur of agreement to disguise its true nature as an expression of doubt. Dale's right that they'll all be painted as heroes if they successfully effect a rescue, and the thought of the spectacle she's been forced to witness in recent days has her impatient to go kicking in some doors, but she can't bring herself to believe that they're going to get Darcourt so easily. To echo the note of caution Meilis sounded, he wouldn't even have to be there to control his squalid little show: everything could be remotely operated, right down to Evian and Mars-bar dispensers attached to each cell door's sealed hatch.

'He might know a bit about computers,' Dale had said, 'but the bottom line is, if our geeks are smarter than him, he loses.'

Angelique couldn't argue with the logic, but nor could she imagine Darcourt missing it either. He had taken not a few risks to play out such a high-profile pantomime, but would he stake so much on avoiding electronic detection, and on the cops not calling his bluff with regard to his ability to intercept their traces?

On the other hand, the arrogant fucker loved showing off, and it was his complacency and his desire to shove it in Raymond Ash's face that had brought about his downfall last time.

'You set?' Dale asks.

Angelique thinks about three adolescents gasping like landed fish, surrogate outlets for one narcissistic, self-obsessed psychopath's jealous rage. That's all that matters at this stage. If they have to play another game tomorrow, she'll have a hot cup of coffee and suit up for that one too.

'Good to go,' she confirms.

'Meilis, how's it looking inside?'

Meilis' voice comes through Angelique's headset after the briefest delay.

'Hostages are all awake, all lying down, all safely positioned away from the doors.'

Dale gives the order: 'Go, go, go.'

They emerge from cover like they've been teleported, swiftly traversing the grass and concrete, closing around the unlet and unlit building like a tightening knot. To the front, there is a blue-painted door at the left-hand side, and ten yards to the right of it a tilting aluminium affair large enough to permit vehicular access to a high-top van. Perfect for getting bodies in and out unobserved. Angelique is heading for the rear, her own request, wanting to be the person on-site if Darcourt tries to bail out the back door.

She is less than twenty feet from contact when Meilis' voice breaks in urgently across her radio headset.

'We just lost two of the feeds, both gone black. Sally and Anika, just Wilson remaining.'

'Fuck,' says Dale.

'Do we hold?' Angelique asks.

Dale takes a beat, a moment so silent and suspended it's like his radio has died. 'Get in there,' he states. 'Get in there fast.'

Angelique signals to the two cops behind her, who are bearing a battering column. As they heft it for their first swing, they can already hear the impact of its counterpart around the other side.

'Got video back,' Meilis reports. 'Both feeds. Each showing . . . I think it's doors. Wait. Could both be the same door. Light is pretty dim. Can't make out what I'm looking at. Oh fuck.'

'What?' demands Dale angrily.

'I'm looking at you. I mean, whoever just came through the door.'

'Are you in? Who's inside?' Dale asks.

Angelique's door buckles and suddenly swings

in as though sucked by a vacuum.

'McGuire, sir,' reports a voice. 'I'm inside. Place is empty.'

Angelique walks through the gap, gun drawn, sees a deserted warehouse containing only more cops and, on a small table in the centre of the floor, a laptop computer.

'Oh, Christ, what is this?' Meilis says. 'It's freeze-framed, both feeds.'

'We can see,' Angelique reports.

The laptop, she spots, has a built-in webcam embedded in the frame just above the top of the screen, pointing at the front door. On the screen are three windows, side-by-side: the live feed of Wilson in his cell flanked by two frozen images of officer McGuire's face as he came through the door.

Darcourt's manipulated voice plays, tinny but distinct, over the laptop's integrated speakers. Angelique is so used to hearing it that it takes a moment to register the significance of it being disguised: this isn't live. He knew they were on their way, the only question is for how long.

'What a capricious mistress is fame. Here we have one man who has been striving to achieve it, and though he doesn't yet know it, he's just been killed by someone who did not seek fame but will now have it thrust upon him. I do not know his name, but by tomorrow everyone in the country will: the unfortunate police officer, following the digital trace I expressly forbade, who was first to open the door and unknowingly trip the wire attached to my decoy computer.'

Angelique looks to the side of the laptop and for the first time notices a second cable other than

232

that leading to the power supply. She follows the fine black-insulated trail along the dusty concrete and sees that it is taped to the inside of the front door, terminating close to the top in a dangling, disconnected contact. At a corresponding height on the doorframe is a small plastic box, on the outside of which a red light is blinking.

'The detachment of this wire has now triggered the release of the lethal gas contained in the canister on Wilson's cell wall. What? You didn't really think I'd keep the oxygen in there, did you? Far too small. Wilson will die quickly and painlessly, as this was not his fault. And after that, the show must go on.'

Angelique watches Wilson's eyes close. He just looks like he's nodding off to sleep. She looks at McGuire, who is physically shaking with revulsion.

'Fuck's sake, somebody turn it off,' says the cop who came in at McGuire's back. He takes a step forward, but Angelique puts up a hand to order stop.

'Nobody touch it,' she says, though she wants to boot the thing against a wall.

The two flanking images of McGuire's face suddenly fade to black, and are replaced with live views of Anika's and Sally's cells. They both remain curled up on their blankets, showing no indication of knowing what has just taken place.

'The canisters in all three cells contain this gas,' Darcourt resumes, 'rigged for release if anybody attempts to open the cell doors. The police were warned, and I will warn them again: do not attempt any further traces. You gambled once. The numbers do not stack up in your favour.'

Angelique thinks he means the odds, but then

233

the view of Wilson's cell disappears, replaced by two sequences of digits. The second is today's date, the first, she assumes, Wilson's date of birth. The figures remain on view for only a few seconds, then reorder and change until a single sequence appears in their place: twelve digits divided into groups of three.

'Oh, fucksocks,' Meilis grunts.

'What?' asks Dale.

'The numbers do not stack up in your favour,' Meilis repeats. 'The cunt just quoted us our own IP address.'

*   *   *

A pub in Soho, appropriate given that they now appear to be in the entertainment business. A group of them retreated there after the debriefing in an act of solidarity, wagon-circling and mutual wound-licking. The press conference was particularly horrible, but Dale acquitted himself admirably. He not only refused to name McGuire, but threatened to exclude any media outlets who subsequently dug around and published his identity. Everybody—McGuire most of all—knew it would be only a matter of time before it was public knowledge by one route or another, but the message it sent out was important.

McGuire's face is on the front of the *Evening Standard*, a split-frame image with him on the left and Wilson Gartside on the right. It's everywhere: she can see four copies from where she's sitting in the pub.

'Only serving the public interest, I'm sure they're telling themselves,' Angelique remarks.

'What can you expect,' Dale replies. 'It's the metropolitan evening edition of the *Daily Mail*: by cunts, for cunts.'

Someone at an adjacent table has the paper open. An inside headline reports that Anika is now the bookies' favourite to win. Angelique feels a shudder when she catches herself wondering what the prize might be.

It's just her and Dale who remain. The ones with wives, husbands, partners didn't dwell, headed off to the people they really wanted consolation from. Leaving only those with nobody to go home to.

Dale mentions stories he's heard about her from Shaw. It's a prompt, a kind of cop therapy. You remind yourself of the successes, all your personal myths and legends, so you can pick yourself up after a bitch like today and face this shit all over again tomorrow. He smiles for the benefit of their mutual morale, but there is a sadness in his eyes. She's seen it there all evening, same as she's noticed him swallowing just a little too often, his face contorting mid-sentence as he disguises his welling emotion. He's disguising it, but he's still feeling it, feeling it all. He's different that way: she's seen way too many who've become emotionally cauterised by his stage. They still do the job, but they've long since ceased letting themselves feel what Dale still dares to. He's a decent man, a good man. There's a lot they recognise in each other.

He asks if she wants to grab a bite. They both know it means all kinds of other things. She knows he could just be looking for a soft place to land after today, but then right now maybe she is too.

She's about to assent and finish her drink when her phone sounds, signalling a text. She glances at the screen, sees the hypnotic, compelling sight of an unrecognised number. A moment after luxuriating in off-duty thoughts of going off with Dale and 'seeing what happens', she is reminded that she's never really off-duty. She has to take every call, read every message.

She opens the text.

If you are not alone, excuse yourself. You are about to receive a message that will elicit a strong emotional response, which you will not wish to have witnessed by anyone who knows you, in particular the police.

Angelique has got to her feet automatically, feeling her pulse increase, her adrenaline begin to flow. She looks at Dale. 'I have to make a call, gotta go outside,' she says. 'Too noisy,' she adds, unnecessarily proffering an explanation.

Dale nods acknowledgement and gestures that he's off to the gents.

Angelique hears the phone chime again as she walks through the pub's outside door on to the Soho pavement. This time there's a picture attachment.

This photo of your parents was taken half an hour ago. There was no holiday. Understand this: your parents will die in unimaginable pain—they will be literally tortured to death, which we will record and send to you as proof—unless you deliver to us, alive and untainted by the authorities, our mutual quarry, Mr Simon Darcourt.

Angelique opens her mouth, but no words issue.

If the police learn of your predicament, they will take you off the case. You cannot tell anyone. Do not be naive enough to believe there is anybody you can trust to share this. We have sources who do not even know they are sources. You are our agent now: understand this. Accept it. Deliver Darcourt or your parents will die screaming.

Angelique stands staring at the handset, facing the tiny but devastating image of her mum and dad, mouths gagged, eyes fearful.

There isn't even time to reel from this, because through the pub window she can see Dale emerging from the gents and looking for her at their table. She runs to the edge of the pavement, hails a passing Hackney and climbs in, telling the guy to just drive. He's about to query this until he catches the look on her face, and pulls away. It takes her a hundred yards or so to compose herself sufficiently to tell him the address of the serviced apartments, and she has no sooner done so than she has to tell him to pull over again. She opens the door of the cab and is sick into the gutter.

The driver mutters something about 'one too many' and thanks her for not spewing up in his cab. He sounds miles away, in some other world, the parallel dimension she was living in five minutes ago but to which she can now never return.

237

She feels both instantly sober and instantly accelerated into hangover symptoms: she can just about think straight, but every thought makes her head spin and makes her feel she might be sick again. She's rationalising, triaging the threats.

Dale: she was on the verge of leaving with him, now he's the last person she can afford to make suspicious of her behaviour. She can discount her exit tonight, at least, given the circumstances. He'll think she just bailed: suddenly decided what they were about to get into was a bad idea. There would be no questions probing further into that.

Whoever did this knows nobody will report her parents missing. They've told everyone they're off round the world for three months. There's James, though. How often did he speak to Mum and Dad when they were on holiday? She had no idea. She'd have to make something up, tell him they'd called her to say they'd had their phones pinched. She had to protect him from the truth.

Like the text said, there was no-one she could tell, especially not her fellow cops. She had to conceal her predicament, as her best chance of keeping her parents alive was to stay on the case. And if she could somehow manage that, all she then had to do was unilaterally apprehend a target the police's massed efforts had thus far been unable to even locate, before betraying all of her colleagues and delivering her prize into the hands of some anonymous but highly organised and evidently well-informed criminal conspiracy.

Impossible.

That's the word that keeps sounding in her mind as the taxi weaves its way through the Soho streets.

Impossible.

She feels the well of coming tears, knows she has to stem it. Has to fend off approaching despair, a collapse she might never pick herself up from.

She takes refuge in cold logic. Whoever has her parents isn't buying a fucking raffle ticket here: they wouldn't go to this bother if they didn't think she could pull it off. Unless, of course, her inevitable failure is part of some greater strategy, but she can't afford to consider that. It was daunting enough just attempting to contemplate what kind of elaborate human-being-sized version of the cup-and-balls trick she'd have to effect in order to deliver Darcourt *and* ensure that whoever was behind this didn't simply kill her and her parents once they had him. And this, of course, would only be *after* she had accomplished the magical feat of making Darcourt, if he was ever found, vanish from under the noses of everyone else who was looking for him.

This wasn't a job for a cop, it was a job for a conjuror.

West End neon flashes past outside the windows. She recalls another late-night cab journey in a different city, and remembers the man who was secretly waiting for her at the end of it.

'Alakazammy,' she whispers to herself.

BAR ACT (II)

Albert's got a seat near the front: got in early and was rewarded with a little table off to the right of the stage. Let the waitress sort him out with a serviceable Martini and got himself comfy to enjoy the show. Eyes on the prize, that's the first thing to remember. The second, and only microscopically less crucial, is to make sure the prize doesn't have eyes on you. It's a right weird feeling, when you're surveilling somebody who doesn't have an earthly that he's being watched: it's like you're looking at them through a two-way mirror, they're right in front of you and yet utterly oblivious. But you need to keep your nerve, which is harder than some folks might think, because some little cautious part of you just can't fucking believe they ain't tumbled you, can't believe they don't know you're there and aren't seeing you as clearly as you're seeing them. That's when you can get jumpy and give yourself away, and it don't take much. It's amazing how the slightest extra effort to be inconspicuous can be the single most conspicuous act. Nah. You gotta believe in yourself, have faith in the craftsmanship with which you constructed that imaginary mirror.

Back in Palma, he had been watching that Innez bastard for days without the cunt having any inkling he was there. Sat in his bar, observed him like a specimen under a bloody slide. First he knew about it was when he walloped him with Mr Spank. Fat lot of fucking good it had done Albert in the end, but you live and learn, don't you? And the better you learn, the better you live. Had to be more careful this time, had to apply a bit of the old subtlety and finesse, now that the quarry knew what he looked like. Hence the mufti: the fake Barnet, the Gregories, the face-fuzz and the middle-aged clobber. His own mother wouldn't recognise him. Well, actually, truth be told, she might detect a certain unnerving familiarity, given that when he'd clocked himself in the mirror before leaving the cabin, the image in the frame was disquietingly reminiscent of his late uncle Vic, as captured forever in his pulling attire in a photo that had pride of place on his dear departed Nan's mantelpiece. Point was, though, Innez, who'd only glimpsed him for a matter of minutes, wouldn't know him from Adam, hence he could sit sipping his Martini less than ten feet from the stage with the same invisibility and thorough impunity he had enjoyed watching the lad's smaller-scale magic act back in that boozer in Palma.

Oh yeah, Innez has gone up in the world, rapid ascent too. A few weeks back, he's turning bar-top card tricks in between pouring beers and washing bottles. Now he's got himself a proper stage and, by the looks of it, pretty bloody close to a full house.

The houselights go down and the stage is lit with a single spot, picking out a four-legged table close to the front. A voice over the speakers asks the audience to welcome 'Maximilian' and Innez walks on. Got a round of applause already and the geezer hasn't done nothing yet. He's scrubbed up, got himself a jacket and a top hat and he's carrying a suitcase. Looks like some Edwardian doctor paying a house call, apart from the peroxide job visible just under the headgear. He sticks the case on top of the table and flips it open, pulling out an oversize black die, also unsettlingly reminiscent of Albert's uncle Vic in that he used to have two such matching items dangling from the inside rear window of his much-cherished conveyance, the superseded but never truly replaced Ford Cortina.

He holds the die up in one hand for a minute and parks the case out of the way on the floor in front of the table. It's just occurred to Albert that a stage magician in this classic get-up is usually accompanied by a nice piece of crumpet dressed in something that would threaten imminent arrest or possibly just hypothermia if worn outdoors, when Innez, or Maximilian, announces that his assistant has only taken the hump and hidden herself inside the die. He gives it a little shake and holds it up to his ear, doesn't hear nothing but says he'd best not shake it again in case he really burns his bridges. Albert can't see where this is going. Gotta be a distraction for something else. He's going to throw it across the stage and she'll appear from the wings suddenly to catch it, something like that. Fuck knows.

But then the thing starts to grow, like some speeded-up time-lapse film of a plant. Soon enough it's so big he needs both hands to hold it, and it's still bloody growing. He's struggling under the weight, so he rests it on the table, where it continues to expand, all of this accompanied by weird, hypnotic music on the speakers. The music stops and the die seems to have ceased its metamorphosis too, though it gives a little shiver when Innez goes to touch it, which makes everybody laugh. He backs off, then makes to approach it once again, more cautiously this time.

242

His hands are just about to touch either side when Albert, like half the bleeding room, jumps in his seat as the aforementioned crumpet bursts out the top of the thing where it sits on the table. Albert is a little disappointed to observe that she's in a suit very similar to Innez's (except for it being covered in white spots, like the die), as opposed to some item of bathing attire you could comfortably fit inside a cigar tube and still have room for the cigar, but that aside, like everyone else in the lounge, he's impressed.

Innez is a clever bastard and no mistake. Found that out the hard way. But he's not as smart as he thinks he is; or rather, he makes the common assumption on the part of clever bastards that everyone else is fucking stupid. A lot of the time he's probably on the money with that, let's be honest, but not when it's Albert Samuel Fleet. Just because he pulled a right nifty move on him once before didn't mean all future engagements were a foregone conclusion. You live and learn. Case in point being the very reason Albert is here on this tub. Oh yeah. Innez knew a thing or two about misdirection, give him that, but short-term gambits could have a nasty habit of backfiring if you used them against those who were used to playing the long game. Innez had sold him a nice little dummy the way he'd emptied his cash, made him think he'd no interest in the other contents of Albert's wallet. It was only when the statement came in that he discovered he was actually lighter to the tune of just north of a grand, paid out as a one-off to some travel firm in Palma. Slippy bastard hadn't sunk the boot in hooligan-style, though: that one single item was all he'd charged. Probably wanted clean away first and foremost, rather than risk triggering an authorisation request that would put Albert back on his tail sharpish. And he *would* have gotten away clean if he'd just hopped on a plane, but the charge to Albert's credit card was for a berth on board a cruise-liner.

Could have been another bit of misdirection, of course: send Albert chasing a bloody boat to the Canaries while he has it away on his toes somewhere else entirely. Geezer knew that he'd seen the air ticket to Paris in his bedroom, didn't he? Could have been another cheeky little play that Albert would have needed to weigh up before making his next, potentially expensive move. Yeah, could have, if it weren't for the fact that he only discovered this lead after ten days of fruitless pissing about in naffing Paris.

Nah, something about this cruise thing just smelled right. The guy wanted to lie low, drop right off the map for a while now he knew someone was on to him. According to the travel outfit, that boat sailed the very morning he'd escaped from Albert's handcuffs. Innez had mumbled something about it being tricky getting him off an island. How much harder did Innez perhaps reckon it would be to get him off his own little *moving* island?

Not as tricky as you might think, my son, especially with the able assistance of Mr Spank's soft-footed little cousin, Dr Rohypnol. Don't you worry about that. Maximilian's would be a limited engagement, shame to say, but as befits the man of mystery, he would finish with a vanishing act and leave them wanting more. Bit of a comedown for a VIP to be disembarking via the luggage ramp rather than the gangway, but he wouldn't be the first star to require an incognito departure route.

He's on to card tricks now. Same ones he was doing in the bar in Palma, mostly, but they're going down just as well. Geezer's got the front for it, and that's half the battle. The thought of the boozer in Palma reminds Albert his glass is empty, and he don't half fancy a refill. There's waitresses working the lounge but he can't get anyone's attention: seems his choice of a table so near the stage has its downside, as they don't want to be moving back and forth right in front of the show. Bollocks. He can't get up and go to the bar; same as the waitresses, he doesn't want to draw attention to himself, though for a very different reason.

Then it turns out Maximilian only reads naffing minds as well, don't he?

He walks across to the left-hand side of the stage, where there's an area curtained off, and pulls a cord. The curtain swishes clear to reveal a narrow cocktail bar, padded with studded leather and bearing a row of stemmed glasses along the top, as well as a translucent cylinder. It's a lot like the ones every nouveau-riche punter used to have in their living room, intending to create the impression of upper-class sophistication but paradoxically serving instead to reinforce the fact that they was one generation out of a room-and-kitchen in the East End. Needless to say, Uncle Vic had one. If the cylinder turns out to be a lava lamp, then that would just put the tin lid on it and he'd be right back at Vic's place in Ilford. Albert remembers he raided Vic's bar one curious Sunday afternoon at the age of nine and got his first taste of gin, with the result that his next taste of the stuff was close to two decades later. Quite likes it these days, mind, and would sincerely like another Martini right now. Fat chance of that though, he reckons, and he ain't the only one sitting there thirsty.

Maximilian reaches into the bar and pulls out a bottle of vino, which he then lobs over his head, letting it spin a couple of full rotations before catching it behind his back in his other hand. He twists off the screwtop, saying, 'Thank God for the Australians—kinda slowed the pace of the trick when I had to uncork it,' and grabs a glass from the top of the bar, holding it by the stem. He pours out a glass of vin rouge and has a sniff before sampling a brief taste, then wrinkles his hooter like he ain't impressed.

'It's chilled,' he says. 'Really ought to be room temperature.'

He holds the glass in front, about eye-level, then waves his other hand, wiggling his fingers like he's casting a spell. The wine bursts into flames: seriously, real fire, not just a little blue tickle like on top of a Sambuca.

'Sorry, overdid that a little,' he says. He goes back to the bar and places the bottle back down out of sight, lifting a chrome cocktail shaker in its place and pouring the still flaming wine into it. He holds it up in one hand and gives it a bit of a swirl. Meanwhile the assistant has entered from the other side with this huge glass tankard, supporting it in both hands, as it looks like it must hold two pints at least.

'Maybe something cooler, huh? How about a Margarita,' he suggests, and pours the contents of the shaker into the tankard. Instead of red wine, it's a clear green liquid that pours from the shaker.

246

And pours, and pours, until the tankard is close to full, despite the shaker being less than half the size of the thing. He places the shaker back behind the bar and takes the tankard from the assistant.

'Kind of overdid it again. Who's thirsty?' he asks, to plenty of giggles and not a few raised hands in the audience. 'No, wait a second, it's pretty dark down there. Better make sure I can see what I'm doing.' With this, he flips a switch and the translucent cylinder reveals itself to be a white lamp. 'That's better. Wouldn't want to spill this,' he says, picking up the shaker again and carefully pouring the contents of the tankard into it. Once again, it appears he bought it from whoever built the TARDIS.

'This is a phenomenon which you'll encounter in any bar,' Innez tells his audience as the Margarita continues to implausibly flow into the shaker. 'Usually applying to the optics, so that what looks like a lot of booze as it's pouring out, somehow becomes that tiny measure at the bottom of your shot glass.' Everyone laughs, Albert especially. It's bloody true of every boozer he's drunk in.

Innez finishes pouring and the assistant puts a cap on the shaker before taking the tankard and exiting stage left. 'I love a Margarita,' he says, walking forward and contemplating the shaker. 'I feel it infuses the whole atmosphere with its ambience, even the light itself.' As he says this, he gesticulates towards the bar with the shaker, at which the white light suddenly turns green. 'I wonder how it does that. Probably just the booze.'

He wanders back to the bar, handing the shaker to the returning assistant en route, then switches off the light and removes the translucent shade.

Inside there's a huge bulb connected to the socket, which looks at first to be green. However, once he's unscrewed it and held it up, Albert can see it is actually clear, and filled with green liquid. With a flick of the wrist, Innez removes the brass end from the bulb and pours the liquid into three of the stemmed glasses on top of the bar. He puts them on a silver tray and carries them to the front, where he invites anyone who fancies one to come and get it.

There's a woman four tables along who steams right in there for her and her mates, who've already emptied a jug of Margaritas between them and are having the same luck as Albert getting a refill.

'How is it, ladies?' he asks. 'The real deal?'

'Better than from the bar,' one of them reports, and they all get stuck in.

Innez walks back to the centre of the stage and takes the shaker back from the assistant. That's when Albert's sussed it. It's the shaker, innit? Got to be. Innez has switched it, several times probably. Yeah. Albert's always been good at tippling magic tricks. He'll have his eye on that shaker now.

'Now, you may be wondering whether the effect I just mentioned works both ways, given the capacity of that bulb is smaller than the capacity of my shaker. Truth is, it doesn't. I've got some Margarita left in here. But maybe that's not everybody's cup of tea. Or Long Island Iced Tea, even.'

Yeah, here we go. He goes back to the bar and pours out a brown-coloured drink into a stemmed glass. Switched the shakers, definitely. Smart, but not as smart as he thinks. That's why Albert's gonna have him.

Innez brings the Long Island Iced Tea out to a keen recipient in the audience. All eyes are on this bloke as he takes a sip, waiting to see if he'll verify that the drink is what it purports to be. All eyes, that is, except Albert's, which remain locked on the shaker in Innez's hand. The cocktail approvingly appraised, Innez walks back to the bar with applause ringing around him. Albert keeps eyes on, waiting for the switch, but Innez stays in front of the bar in full view as he lifts another glass and suggests: 'How about we move from Long Island to Manhattan?' He puts a cap on the shaker and gives it a thorough jiggle before pouring out a distinctly amber-coloured drink. The assistant comes over again, this time bearing a tray of fresh glasses, upon which Innez places the new drink and the cap.

All right, so it's the lid, Albert decides, but Innez immediately takes another stemmed glass and pours a clear drink into it from the same shaker. Just to rub it in, it's only a bleeding Martini, innit?

Holding the Martini in one hand, Innez leaves the shaker on top of the bar and lifts the Manhattan from the tray, then takes the drinks down to the audience. He asks for opinions from the grateful recipients. The woman who gets the Martini says it's the best she's had since they left Marbella, asks him if it's Tanqueray. Albert's throat's like an Arab's sandal just looking at her knocking it back.

He returns to the bar, where the shaker remains in plain sight, the assistant still standing there with her tray of clear glasses, and now it just gets daft.

He pours out, from this same bloody shaker, a cherry liqueur, a Crème de Menthe, a Piña Colada, a Bloody Mary and a Blue Lagoon. Albert's all but licking his lips as he watches the assistant carry the tray down to the floor and start handing out the goodies. Innez has walked to the front of the stage to take the applause, still holding the shaker. He has a look back and forth and suddenly zeroes in on Albert's empty glass.

'You ready for a refill there, sir?' he asks.

Albert thinks for a split-second about waving the offer away, keeping his head down, but remembers his own rule of keeping your nerve: the slightest extra effort to be inconspicuous can be the single most conspicuous act. Added to that, the guy can't possibly see any of the audience's faces with much clarity while he's on stage under the lights and they're sitting down in the dark. And all of this is to say nothing regarding precisely how much Albert fancies another drink.

'Wouldn't say no,' he replies, deepening his natural voice a little and neutralising his accent.

'Elizabeth, bring me that man's glass.'

Now, this really bloody surprises him. He'd assumed Innez would just toddle back to the bar for another glass as a pretext for some nifty bit of legerdemain, but it is indeed the glass he so long ago finished, and this glass only that is presented to the magician on her tray. Remaining at the front of the stage, he takes it in his hand, gives it a sniff.

'Martini?' he asks.

Albert nods. Innez places his glass back on the tray and reattaches the cap to the shaker before giving it a good old once-over. He then pours out a fresh drink, complete with an olive, and walks down to present it to Albert, with the audience giving it plenty. Innez hands it over with a smile, looking him right in the face without the merest glimmer of anything that could be interpreted by the most cautious mind as the slightest hint of recognition.

'How is that, sir?' he asks.

Albert takes a mouthful. It's chilled, it's smooth, and it's a sight better than the one he got from the bar.

'Delicious,' he declares, then watches Innez return to the stage, lapping up the ovation before one last surprise to close the bar act: pouring himself a drink from the shaker, which turns out to be milk.

Albert rests back in his seat to enjoy the rest of his drink and the rest of the show. No doubt about it, it's a bloody good Martini. Bloody good routine, too. Almost a shame to deny subsequent audiences the unquestionable pleasure, but not everybody could get paid for doing card tricks and stunts with a magic cocktail shaker. Nothing personal. And he wouldn't be taking any chances this time, wasn't going to be underestimating a bloke like Innez. You live and learn.

The assistant rolls away the bar, then high-tempo drum music plays over the speakers as she performs a quick change. She returns dressed in a black leather outfit with understated but inescapable S&M overtones, bearing two armfuls of handcuffs, padlocks and chains. She and Innez perform a bit of a dance to the rhythm for a minute, establishing that she's the one wearing the strap-on dildo in this particular relationship, then she leads him over to the pillar near the right-hand side of the stage. As part of the dance, she commands him to his knees in front of the column, then pulls his hands behind his back either side of it and cuffs them together.

Albert feels the old ticker give a start, but reminds himself it don't mean nothing. What happened back in Palma must've given him the idea, that's all. The girl cuffs Innez's feet together too, at which he feels a little light-headed. Tells himself not to panic: it's the old *déjà vu* all over again, got him a bit dizzy, innit.

The assistant rolls a screen across from stage left, an opaque drape to conceal whatever technique Innez uses to get out of the cuffs, and Albert would have to confess to deep disappointment and discomfiture at this impending absence of revelation. But that's bugger-all compared to how he feels when, just before the screen obscures him, Innez looks straight at Albert and gives him a wink.

Something inside him turns to ice. Suddenly he feels like the audience has disappeared and he and Innez are the only ones left in the place. The tempo of the music increases, but it's like he can barely hear it, like it's fading out. He grabs his drink and downs the rest of it. After that, everything starts to go swimmy.

*　　　*　　　*

Zal's hands are cuffed, tight about both wrists, the jangling links of solid steel looped behind his back around an upright column. He is on his knees, his bare feet similarly bound to the immovable pillar. The steel is warm from the hands that fastened it, moist now with two people's sweat. His arms are stretched behind him, his back tight to the column, his posture cramped and contorted.

His captor has retreated from sight. Zal is now isolated, hidden from any observer, cut off from all intercession. The bounty hunter, though unseen, remains mere yards away, rapt in his vigil.

Zal allows himself a moment to contemplate precisely how his situation must look from that bastard's point of view: one man relishing the other's inescapable captivity, blissfully unaware that he has the picture back to front.

Zal smiles and whispers to himself: 'Alakazammy, stairheid rammy. Suffer, you prick.'

The escape itself is not, he would concede, much of a spectacle, and he had no intention of making it a regular part of the repertoire: it was intended to be for one night only, and for one spectator only. However, Lizzie had the idea of turning it into a dance, and it seems to have played pretty well, so it may end up being reprised and probably built upon. The whole show is coming together nicely, some genuine finesse now augmenting the mere energy, pace and enthusiasm that got them through some seat-of-the-pants early performances.

Zal had been practising for up to fifteen hours a day, Lizzie not much fewer in the run-up to that first unsanctioned show, with Morrit just as busy behind the scenes. Henderson had attended, as predicted, but earlier than anticipated, turning up to investigate why there was a large crowd queuing outside what was scheduled to be an empty lounge. He wasn't foolish or officious enough to stand between the unsolicited act and an eager audience, especially on a rainy day, though he did tell Zal sternly that he wouldn't be paying anyone for it. To this Zal responded by reminding him that he already had. Understanding that there was very little to lose, Henderson took a seat along with everyone else.

The next day's show was scheduled and official.

They got the benefit of it now being advertised as part of the entertainment programme, but Zal considered it more effective (as well as good practice) to preview the act by walking the decks and bars, performing card and coin magic on the spot. It was kind of like David Blaine, though without the tricks coming across as a cry for help. Also as predicted, the act soon got moved to the bigger lounge, where the trapdoor under the stage allowed them to open with the Expanding Die. Mirrors fixed between the legs either side of the table, plus some careful work with Morrit's tricked suitcase, concealed Lizzie's elevation, assisted by all eyes being drawn to the die itself. DeKolta's original was a complex construction of telescoping brass tubes, a maintenance nightmare and said to require two men to recompress after each performance, but Morrit's reconstruction had the benefits of lightweight aluminium, which led the old man to speculate wistfully about what his idol might have achieved given access to modern materials.

Zal undoes the handcuffs, taking longer about it than he did back in Palma, or at least appearing to. On stage, you never want an escape to seem too easy. He wanted to give Fleet some time for things to sink in too, before he lapsed into unconsciousness.

254

It was often said that conjuring could lay claim to being the world's second oldest profession. It had been around as long as human civilisation, and in those aeons taken innumerable forms, but there were certain fundamentals of it that never changed. In 1634, a London writer calling himself Hocus Pocus (believed to be one Samuel Rid) published *The Anatomy of Legerdemain: The Art of Jugling* [sic], in which he prescribed these essential requirements of one wishing to practise the arts of 'conveyance':

First, he must be one of impudent and audacious spirit, so that he may set a good face upon the matter.
Secondly, he must have a nimble and cleanly conveyance.
Thirdly, he must have strange terms, and emphatical words, to grace and adorn his actions, and the more, to astonish the beholders.
Fourthly and lastly, such gestures of body as may lead away the spectators' eyes from a strict and diligent beholding of his manner of conveyance.

In 1716, one Richard Never added:

He must have none of his trinkets wanting when he is to use them, lest he be put to a *non-plus*.

This last was the one most solemnly imparted to him by his father, and remained the touchstone that served his every endeavour. Zal had been preparing the stage for Albert Samuel Fleet's arrival since long before the guy even set foot on board the *Spirit of Athene*. Strictly speaking, Zal's preparations for their second encounter had begun before *he* set foot on the ship.

Zal had believed Fleet unquestioningly when he said he'd find him again. That was why, once he'd opted for travelling hopefully aboard the cruise ship, he decided to use the bounty hunter's credit card to pay for it. He didn't need the guy's money, merely needed him to pay for his passage, so that Fleet would know where to find him, and Zal would know he was coming.

Once he had established his credentials as a valued member of the entertainment crew, Zal was soon able to charm the ship's purser into setting an alert for Fleet's name against the advance passenger manifest, knowing he'd have to embark under his real name because he'd need his passport. This ensured that when inevitably he got his monthly statement, joined the dots and made his booking, Zal knew not only when and where he would be boarding, but right down to which cabin he would be assigned also.

Fleet joined the ship in Tenerife, having flown out there to catch up with it. Zal watched him board from an overlooking deck. His disguise was pretty good, but only if you weren't specifically on the look-out for the sonofabitch and didn't already know where he was going to be.

Swiping a passcard from housekeeping, Zal turned over his cabin while he was having lunch in one of the restaurants. It appeared Fleet had invested in some new cuffs and an outsize suitcase with wheels. He hadn't brought anything to fill the

256

thing, however, as the intended contents were standing looking at it. Zal also found several phials of Rohypnol, which was how Fleet was planning to get him into the big luggage and on to dry land. Either that or he just placed little faith in his chat-up lines.

He knew Fleet would come to the show, same as he'd kept an eye on his prey at the Dracon Rojo. Zal watched him come in from through the curtains and enlisted a friendly waitress to let him know what he was drinking.

Zal remembered his dad trying to work out how Alan Wakeling did his ingenious bar act, a quest that had continued after Wakeling passed the apparatus and techniques on to Earl Nelson. Wakeling's itself was merely the latest improvement upon a trick called the Inexhaustible Bottle, dating from the early nineteenth century: like everything in magic, its evolution was ongoing and its secrets plundered. Soon enough, his dad incorporated a version into his own repertoire, and around that time variations of the act became common. As a young child, Zal had been uninterested in the mechanics of it, just dazzled by the impossibilities and the cascade of different colours emerging from the shaker. He had confidence that such a spectacle would still entertain an adult audience today, and giving a few of them free drinks never hurt your popularity either.

It was all in the routining. His dad had puzzled over it, constantly changing his mind about whether it must be done by switching the shaker, by use of imitation drinks or by some kind of chemical solution. The switch theory fell down on the sheer number of such transpositions that would be necessary, while the other two collapsed to the sound of approving thanks from all those satisfied customers. Like all the best routines, of course, the effect was not achieved by a solitary technique, but by a manic combination of all three. There were hollow-stemmed glasses involved, a cleverly constructed drinks tray that concealed what cocktail constituents were already at the bottom of hi-ball tumblers, as well as shaker caps filled with various liquids too. Most nights, only one drink was unapproved for human consumption, and that was the flaming wine, containing as it did sodium carbonate, phenothalene, potassium and lighter fluid. Zal had added it to the act after they put in at Puerto Del Carmen and he was able to get hold of the more exotic constituents.

The main shaker dispensed only vodka, which being essentially tasteless, worked as a standard base solution (or at least passable substitute) for most of the cocktails. It also gave them all an extra alcoholic kick, which ensured that the customers immediately found their drinks to be the real deal. Tonight, however, there had been one that packed more of a punch than usual, that final shaker lid containing Vermouth, Tanqueray, one olive and a generous dash of Fleet's own sedative.

Cheers.

THE TYRANNY OF NORMAL

She was expecting some dodgy lock-up under a railway arch, a crumbling end terrace with a garden full of fridges, or maybe some semi-derelict rural cottage with rabbit heads impaled on a surrounding barbed-wire fence. Instead she's in a new estate, all burnt-ochre brickwork, double-glazing and monoblocked parking bays. There's a Dora the Explorer trike lying on its side in front of a Toyota SUV and a new-model Honda Civic in the driveway of the address she's been given, with the Bugaboo infant equivalent of a 4×4 all-terrain vehicle sitting at the front door. So, two cars, two kids and a new-build in suburbia. This can't be right. Either the new estate has a duplicate street name for somewhere else entirely or her sat-nav is taking the piss.

Angelique gets out of her car. She has to check it out anyway. There was always going to be an abundance of wrong turns and dead ends on this quest. Tracking down Zal Innez strikes her as only slightly less of an ask than tracking down Simon Darcourt, but at least she has a lead in the case of the former, and it's a good displacement activity to distract her from contemplating the abject absence of any such clue towards the whereabouts of the latter. Minute to minute, she can't decide whether no news is good or bad news on that score. The lack of progress feeds her feelings of impotence, the dread sense that he'll disappear again, never to be found; and yet whenever there's even a whisper of a new lead, she feels sick in case it becomes something momentous, something that accelerates matters beyond her control and delivers a solution that she is powerless to influence. Searching for Zal at least provides the reassurance that there's an aspect of this chaos being driven by her own hand.

* * *

The hardest thing is the everyday. The hardest thing is that she just has to get on with it. Nobody stops all the clocks, nobody dismantles the sun. As people are wont to remark around *someone else's* bereavement, life goes on. When you want the world to stop, when you want the world to acknowledge the enormity of what you are enduring, life indifferently goes on. At least with bereavement, you usually get cut the slack of a few days off work. Angelique's greatest trial is that she still has to show up, put in her shift and act like there's nothing wrong. Fortunately she's had more practice at wearing a mask to fool her colleagues than to fool her suspects, but she preferred it when her big secret was that she was shagging the guy her fellow cops were seeking. There was, at least, a constant exhilaration about that, a sense of never having felt so alive. This time, it's a permanent, grinding sense of grim dread, and the only familiar feelings are the terrible ones: the constant paranoia that everybody *must* be able to see right through you, to read what you're trying to conceal like it's printed on your forehead.

And what's truly, depressingly awful about it is that it's endless: the days pass, she gets on with work, her secret remains just that, and soon enough it starts to feel almost normal. The sick joke is that nothing's changed: she's still doing the same horrible job and if she'd never received those texts, she'd still have been steadfastly avoiding calling her parents. All that's different is she no longer has the option, though as a further twisted taunt, she is sent a new picture of them to her phone every day, usually accompanied by a request for updates on her progress. There has been little danger of her suffering Texter's Thumb in composing her replies.

In practice, it's not difficult to hide her inner turmoil from her colleagues, as a depressed look of resignation and hopelessness is practically uniform these days. Another young victim murdered, a policeman ragdolled by the media machine, and their whole operation donkey-punched by Darcourt. There's so much gloom and despond that she and Dale can barely muster any mutual awkwardness about what (almost) happened between them. Why pick out one particular turd when you're floating on a sea of shit?

There is a growing stench of defeat about the place. It's unspoken but unmistakable: cops know it when they smell it, particularly the more experienced ones. They've all worked cases where they have inescapably come to realise that there can be no winning, only degrees of losing, and with every mistake—indeed every news cycle that passes without progress to report—magnified and scrutinised by this thousand-eyed watcher that never sleeps, this case is the gorgon-headed bitch to top them all.

Every day, in some new aspect, it seems to get worse, seems to hurt that bit more. After the frantic but disastrous search (which at least had felt like they were *doing* something), had come the blame-and-shame bonanza of the aftermath. The siege effect at least had the benefit of pulling everyone together rather than looking for scapegoats or turning on each other, but they *do* feel shame, they do know they've failed, and they've been left not just looking powerless, but humiliatingly chastened as Darcourt's diseased show goes triumphantly on.

With Anika and Sally looking equally endangered, Darcourt announces that the following will be the final twenty-four hours of the competition. This leads to one of his servers crashing without any help from Meilis, as the traffic surges in the lead-up to this appointed but tantalisingly unspecified conclusion. It also leads to a face-off between Dale and the head of the Met himself, Commander Aldwyn Keen, who has ordered that there should be no further attempted computer traces. Dale orders Meilis to disregard this directive and unleash everything his geeks can muster, figuring they have nothing to lose. Meilis, having been placed in an impossible position, refuses the order until it is referred up. Angelique would have to admit to harbouring a slight suspicion that Dale is trying to engineer his own dismissal from the investigation, but when Keen turns up in person to have it out toe-to-toe, there are no histrionics. Dale argues his case, calmly and logically, and he wins.

'We know he isn't bluffing,' Keen states. 'If he detects another trace, he'll just kill one of the girls, and quite probably both.'

'Not today, sir,' Dale argues. 'He's been working

on this for ages, and now it's the climax. He's got some big finale planned, and he's not going to cancel that. There's only two girls left. If we wait until a few hours before his deadline, there's no way he's going to pull the plug on his own big moment by prematurely ending the competition.'

Keen concedes the point, though Meilis predicts his team's efforts won't yield anything anyway.

'Ever since the debacle of the first trace attempt, I've strongly suspected he's got hold of a GOG,' the alpha geek opines. 'And a state-of-the-art one at that.'

'What's a gog?' Angelique asks.

'Gee-Oh-Gee. Hacker-speak again. Stands for gottle o' geer. It's something—can be hardware or software, and I suspect his combines both—that disguises the source of a signal.'

'But if you know what he's using, is there any way of, I don't know, detecting *that*?' Angelique suggests, none too hopefully.

'Not this time, I fear. Used to be like virus and anti-virus apps, a constant arms race. You could retro-engineer a solution as long as you had a copy of the virus code, just as you could develop a new virus if you knew the algorithms being used by the latest detectors. But I've been looking into this since the last trace, shook a few trees. What fell out isn't good news. There's GOGs available now that would allow you to upload from the next room and you wouldn't be able to detect it from where we're sitting, and it doesn't stop with internet feeds. There's signal-shielding hardware out there that would let you transmit all kinds of stuff without giving away your position. It was developed for military application, but it was developed by tech geeks, and in tech geek circles, nothing stays exclusive for long. We're not going to find this bastard by sourcing these feeds, and we'd better hope he doesn't decide to start his own TV station, because if he's got the kind of kit I think he has, we won't be able to find that either.'

In the event, both Meilis *and* Dale are proved right, albeit to no-one's satisfaction. Darcourt doesn't respond to the attempts to trace his signal, and the efforts prove futile anyway. The appointed deadline is reached without Darcourt bringing a premature end to the proceedings. It is only in what happens next that Dale's assessment and prediction fail him.

There is no big finale. There is nothing at all. The feeds simply go blank, and shortly after that, the whole site goes offline. Darcourt doesn't even, to the tangible but unvoiced frustration of the entire UK media, announce who won.

They stare at the blank monitors, waiting with an animated dread for them to kick back into life with some new depravity. Every one of those first few seconds seems to bring closer some inevitable revelation. But the seconds become minutes, the minutes hours and the hours days.

Though Angelique hadn't believed it possible, for about forty-eight hours there seems to be even more coverage of the story than while the *Dying to be Famous* competition was live and online. Rumour and speculation pour out of journalists and TV reporters even more incontinently than the previously endless slurry of comment and redundant description. Darcourt has killed himself and both girls, that's one. Aye, right. Pipped only in Angelique's bitter-smile stakes by the theory that the police have secretly rushed in and shut the whole thing down but are keeping it quiet; this reticence being for reasons undisclosed, certainly reasons the reporters are unable to deduce.

Others suggest the whole thing is a giant hoax, a mammoth publicity stunt facilitated by the widespread complicity of relatives, not to mention some very realistic special effects. Sure. Forget Simon Darcourt: has anyone braced Chris Morris?

It's said that nature abhors a vacuum, but not as much as the media. By day three, the story is off the front pages. Lia Wilby, *Coronation Street* starlet and lads-mag lacy-undies-but-no-nipples photo-shoot regular, announces her engagement to Karl Howard, Man City's heart-throb goalkeeper, and that is more than enough to satisfy the tabloids, particularly during a time when a good-news story seems at a premium. (Angelique suspects it is more than mere serendipity that the announcement is made on the first day that it *wouldn't* get buried way back in the papers, but this job could make one terribly cynical.) As for the broadsheets, well, another MP has crossed the floor, with timing just as fortuitous and therefore as suspect as the Wilby-Howard engagement, and George Bush has bombed somewhere else, continuing to strive for as many last shots on the swings as possible before getting finally booted out of the Pennsylvania Avenue playpark.

Only the *Express* stands firm, with a front page pondering whether it was actually Simon Darcourt who killed Diana, but in their own unique way, even that is an admission that they have nothing new on the story.

From a police point of view, despite being very

likely a mass-murder investigation, it is more immediately now a multiple missing-persons case. The only real crime scene has reopened to clubbers, while the warehouse near Walsall has yielded nothing. Despite Darcourt leaving them in no doubt as to his identity, he hasn't left them any DNA evidence to even prove he'd been there. They have no bodies, no evidence, no witnesses, and no solid leads. Or rather, strictly speaking, *the investigation* has no solid leads. Angelique has a very strong one, sent to her phone a couple of days ago along with the latest e-postcard from hell, but she is keeping it to herself until the timing is right. To that end, this otherwise depressing absence of anything more constructive to be getting on with at least allows her the time to pursue certain other, more private, lines of enquiry.

In looking for a bank robber, she decides that the first route is to follow the money, even if the only clear path is the one leading back to where the money came from in the first place. Knowing where the police investigation had begun and ended, she speaks to the RSGN, whom she rightly anticipated had been less content to let the case go cold.

She meets with a woman named Debbie Holland at the RSGN's head offices in Broadgate. Angelique tells her she is looking back at the case because of a possible overlap with something else she is investigating, but which she is not at liberty to disclose. As nobody outside the force (and very few inside it) knows about Angelique's involvement in the bizarre events of that day on Buchanan Street, she doesn't inform Ms Holland that her knowledge of the robbery extends somewhat beyond a cursory perusal of the police file. As it turns out, it is sufficient merely to be on the force and aware of the case in order to be getting the stink-eye from this particular RSGN executive, and Angelique is made unambiguously aware that as an institution they are still very sore about having their pants pulled down by—literally—a bunch of clowns.

Holland is one of those corporate types who acts as though it was personally her money that was stolen. It reminds Angelique of the discussion she had with Zal only days after the heist, regarding the morality of what he had done.

'RSGN could lose the same amount if one of its stock traders had one beer too many over lunch,' he'd argued. 'There are worse crimes than what happened on Saturday. Worse thefts. I ain't saying it's right, but I'm . . . morally at peace with it.'

Angelique had gone ten rounds with him over the issue back then: pointing out how they'd only pass the loss on to the customer or lay off some more staff. Listening to Holland saddle up her high horse in her plush corner office, kitted in her thousand-quid suit and her shoes worth Angelique's monthly rent, she finds herself wishing Zal had stiffed them for ten times the amount.

'I understand that the bank put forward a substantial reward for information leading to a conviction,' Angelique says, over two paper cups of barely chilled water from the cooler; she figures you don't get offered hot beverages if you work in the public sector. 'Did anything precipitate from that? Because what I'm really interested in here is anything that might have been overlooked, or perhaps the significance of which might only reveal

267

itself now we're further down the line.'

'The RSGN did offer a reward, yes,' Holland confirms. She only ever refers to her employer as 'the RSGN', never merely 'the bank'. She is so on-message, it's like she is on a bonus for mentioning the name a minimum number of times. 'We were forced to, truthfully, by the apparent inertia on the part of the police. We know, through channels the RSGN is not at liberty to disclose [woo, touché], that the police had a suspect, but they refused to tell us his name. You'll no doubt know more about this than I do.'

'In case we miss something by taking it as read, assume I know nothing.'

'Oh, I can certainly do that. When it comes to the police on this case, we at the RSGN have had plenty of practice, after all.'

Angelique patiently endures this dig with a thin smile that says: 'Sticks and stones—just give up the skinny.' The smile may also add 'bitch', but that is for the observer to interpret.

'Whoever this suspect was, he had links to organised crime in the US. The robbery, I am reliably informed, was somehow connected to a major police sting in Glasgow shortly afterwards, involving American and Scottish criminals. Unfortunately, the police seemed far more interested in the fall-out from that than in pursuing our case. We're not wet behind the ears, officer de Xavia, we work in a more cut-throat world than the crooks do, so we at the RSGN had little doubt that some deal or compromise must have been struck.'

'I'd need to be much further up the food chain to know anything about that,' Angelique replies.

'I don't doubt it, and nor do I envy you the task of unravelling the tangled web they probably wove to cover it up. The Procurator Fiscal himself told us that even if the police apprehended this suspect, there was simply not enough evidence to mount a case—not without a confession or recovering the cash from his direct possession. That was why we put up a reward: we knew the official channels had been exhausted. We intended—and we still do intend—to bring a private prosecution if we could track the culprit down, but if we were able to get some of our money back, it would have been a start.'

'And what did you get?'

'The anticipated deluge of cranks, nutters, attention-seekers and opportunists. People effectively buying a raffle ticket in case their stab in the dark could later be connected to the solution and provide them with a claim to the reward. I can give you the file, if you've got a dump truck parked outside.'

'What about the criminal fraternity?'

'I beg your pardon?'

'Well, forgive me if you think I'm impugning your institution's integrity, Ms Holland, but as you informed me you're not wet behind the ears, I'm assuming we can cut the formalities and acknowledge what a reward like this was really aimed at. You said official channels were exhausted. I think we both know what the corollary implies.'

Holland says nothing for a moment, pursing her lips in affected silent indignation. Angelique wonders how long corporate policy has stipulated it appropriate to sustain a postured huff when coughing up a distasteful truth might prove beneficial. Turns out to be about eight seconds.

'There was this one guy,' she says. 'Some kind of mercenary or . . . what would you call it?'

'Bounty hunter?' Angelique suggests.

'Yes, that's it. Said he knew where the suspect was and could tell us his old name, new name and precisely where to find him, but he wouldn't give us anything unless we paid upfront. We said we'd give him *some* money if what he told us led to the suspect's apprehension, and more if it actually led to a conviction. He wasn't prepared to accept those terms. He walked away empty-handed rather than even buy his raffle ticket by giving us the info. We reckoned he was bluffing. If he was a bounty hunter and he knew where this guy was, why didn't he bring him in?'

Angelique could think of a few reasons.

'What was his name?'

'Can't remember,' Holland replies. Angelique feels herself slump inside but maintains her posture. 'Hang on though, I'm sure the card he left is still on file.'

*   *   *

Albert Samuel Fleet, the card said. All the numbers on it were long dead, but, naturally, he had a jacket. Petty crime, moving up to gangland runner and fixer, as well as, significantly, police informant. A proven ability to play both ends against the middle had served as a valuable apprenticeship for striking out on his own in the shadow world of the 'procurer', as his card described it with almost charming euphemism. He

270

had channels of information on either side of the law, and had evidently proven useful to both at various times, on a strictly-no-questions-asked basis.

These last few years, however, the sheet was blank. He was running a private surveillance firm, in partnership with a couple of ex-coppers, and unless her sat-nav really was on the blink, it looked like he'd gone respectable as well as legit.

Angelique sees him shortly after getting out of her vehicle, coming from around the back garden in response to the sound of the car door closing. She'd phoned him to make an appointment—best play it courteous, she figured—so he is expecting her. He's wearing an old West Ham top, flecks of grass dotting his face and a pair of gardening gloves on his hands. A little girl appears at the gate barring the path running behind the building, and is led away by an attractive, olive-skinned woman a good ten years younger than the man of the house.

They exchange polite greetings. He takes one of the gloves off and offers his hand. His smile is boyish but his eyes wary, ready to evade and ready to defend. He's been around the block. She doesn't know what she'll get here, but she's got to give it a shot.

She's embarrassed to feel a pang of jealousy; it's fleeting, just pops into her mind like a fly landing on her face: she bats it away as absurd but the fact that it appeared at all is unsettling. This guy, this one-time ducking-and-diving quintessential dodgy geezer, had the happy marriage and the kids and the house and the garden and that air of contentment. Changed by life, or maybe that just happened to you, maybe that was growing up. You quit chasing rainbows and tilting at windmills and realise what will really make you happy. Angelique has realised it herself, albeit too late, the job already having taken the best of her.

'So what is it I can do for you, my dear?' he asks, all self-confident Cockney charm.

'I thought you might be able to tell me where you last saw Zal Innez,' she replies. She reckons this ought to knock him off his stride a little, but it's she who is almost derailed. She stumbles over the name, her voice threatening to choke. It's the first time she's said those two words aloud in years, and merely speaking them feels like breaking the seal on something highly potent and tightly suppressed.

Fleet eyes her very cautiously for a moment. Perhaps he's thinking about giving her the old 'Don't know what you're talking about' routine as the first gambit in haggling a price, but it's mutually obvious that they both know *exactly* what she's talking about. If she had any doubts at all over whether Fleet was bluffing to the RSGN, they are now dispelled. She might have stumbled over the name, but when she said it, Fleet looked like he was afraid his pocket was being picked. This guy has seen Zal, and no mistake.

He settles for: 'Now what makes you think I might know something like that?'

'You offered his last known alias and whereabouts to the RSGN. That's where I got your name. Let's not waste time, Mr Fleet, I know it's going to cost me and I'm prepared to pay. You went after him, didn't you? But you didn't get him. That's why you offered the bank information rather than offer them the man himself. And that's why you rejected their counter-offer of a reward contingent upon your information leading to his apprehension. You didn't fancy anyone else's chance of bringing him in either.'

Fleet holds up two fingers. Angelique hides her anxiety behind a poker face: she always knew she'd have to negotiate, and with an arch-haggler no less, but she'd told herself she wouldn't go above one thousand and the bugger was starting at two. She had the money, but she didn't know what else she was going to need it for in her secret little quest.

She's about to lie that it's not worth that to her, fearing that this will tell him it's worth *at least* that to her, when he reveals he wasn't indicating a price at all.

'Twice I went after him,' he says. 'First time, I had him handcuffed to a pillar; hands and feet. Thirty seconds later, it's my hands, my feet that's chained to the same pillar. Second time, I didn't fare any better. Not against him, anyway. That's why I didn't fancy the bank's chances, or anyone else's.'

'Who else did you offer this information to?'

'The notorious Mr Bud Hannigan's people put me on to Innez in the first place. But by the time I got back from me travels, as you'll know, Mr Hannigan had changed his titular prefix from "the notorious" to "the late", and his successors had different priorities. Besides, I had different priorities myself by that time. My second encounter with Mr Innez convinced me to get out of that whole game. You live and learn, I always say, and if you don't learn, you might not live.'

'I realise it's going to cost me,' she states, taking his meanderings to be a means of dangling the goods. 'I've got five hundred in cash right here.'

'Listen, darlin', I will happily take your wonga to give you the name he was using and the place I last saw him. But the most valuable thing I could give you would be to keep this shut. That way, I'd be sparing you a lot more than just your money.'

Angelique takes out a roll of notes from her bag. 'I only want to talk to him.'

Fleet pauses to think for a second, then to her surprise accepts the roll.

'I sincerely hope that's true, love,' he says, flipping through the notes to count them. 'Because even if you go after him with the whole of the SAS, he'll find a way of vanishing. Just when you think you've got him, you find you're left holding sand. But then you know that, don't you, because I'm guessing he's vanished from you before.'

Angelique tries not to betray any affirmation, but she'd be as well nodding like a car ornament. They both know.

'See, when he had me handcuffed, I drew on a bit of the old bravado and told him I'd catch up with him again. I am thus compelled to ask you the same question he put to me at that point, which was this: what makes you think it'll be any different next time?'

Angelique feels a bullet to the heart, for this is the question she's been asking herself since the night Zal slipped silently from her flat while she slept.

'I don't have an answer,' she admits. 'But same as you, I need to find out first hand.'

Fleet nods, understanding but regretful, like he knows from his own experience that this is a mistake she has to make for herself, and from which she cannot be dissuaded.

'Second time I went after him, that was what did it for me,' Fleet says, unbidden. 'He was on a cruise ship, working as a magician. Watching him perform, I thought I knew his tricks, thought I could suss him out. Instead, I learned that this guy would always be one step ahead, and I'd never suss him. Course, I only worked that out in retrospect. At the time, I thought he was in my grasp. One minute I was watching him onstage—being handcuffed to a pillar, which was what tipped me off things might be going a bit hooky. Next thing I know—I mean literally, next thing I know—I wake up in a lifeboat in the fucking Atlantic. Scuse the language, my dear, but it still gives me the shivers thinking about it.'

'How long were you adrift?' she asks, wondering how ruthless Zal was prepared to be. She already knows he's killed when he had to.

'Long enough to precipitate a career change, I'll say that much. Nah, he'd hidden a transmitter on the boat, hadn't he? He never abandoned me to the fates, but he did give me a very long dark night of the soul before salvation arrived: as it happened, in the shape of Señorita Yolanda Gomez and her colleagues on the Islas Canarias coastguard.'

Angelique thinks of the olive-skinned woman in the back garden. She gestures with her head towards the house. Fleet grins.

'That's right. Another reason you have to tread cautiously with this geezer: weird things happen around him. You sure you wanna do this? Because I went after him a time-served Jack the Lad, and came back bleedin' married.'

Angelique laughs politely, but his words remind

275

her that she has barely dared envisage what she even hopes for from this mooted reunion.

I could live with that, she thinks.

*   *   *

She's back in the car, allowing herself the respite of a smile as she muses over Zal's choice of stage name, and pondering the likelihood of him still being on board the same cruise liner. Either way, it's more than a start. Even if he's moved on, she knows he was working and she knows where. That means payrolls, that means tax records, and if he's as good as Fleet claimed and has sought opportunities elsewhere, he'll have taken the name with him. Even in her strung-out and tortured state, she finds herself taking a moment to feel happy for him, wishing she was watching him do his show in some parallel world where there was no Simon Darcourt and her parents were still safe at home with nothing more to worry them than their daughter's inexorable submersion in the quicksand of spinsterhood. But the moment fades, the glimpse vanishes, and she wonders what right she has, if Zal is finally happy, to bring all this chaos to his door. She had decided she was too much of a threat to him way back when it was only her job that complicated things: now she was bringing serial killers and organised criminal conspirators to the party. However, she also knew she had little alternative, and that Zal Innez of all people understood the lines you sometimes had to cross in order to protect the people who had once so invaluably cared for you.

She recalls being on holiday, in Mallorca, when

she was five years old. She couldn't swim, but she spent most of the time in the hotel pool propelling herself around, kept afloat by this rubber ring with yellow flowers printed on it. There's still a photo in an album somewhere in her parents' house, showing her standing with it tight around her middle, the pool's surface glinting aqua blue in the background. James could swim at that stage, and he'd spend ages just jumping in at the deep end: jumping in, swimming back to the side, climbing out then jumping in again. She thought it looked like fun, looked easy, and she had her ring to keep her afloat, so she gave it a go, leaping in right behind her brother. She went straight through the ring and down into the depths of the water. She remembers an enduring feeling of fear and panic, her eyes open but seeing only a stinging blur of blue. Then she felt arms around her, drawing her to the surface, hauling her out of the pool. Her dad had dived in, still wearing his shorts, t-shirt, sandals and watch. He hadn't hesitated, just launched himself immediately to her rescue.

The engine is idling as she sits in a traditional London tailback, which at least spares her any fannying about putting on a Bluetooth headset when her mobile rings. It's Dale. And at last, at long fucking last, it's a lead: an eyewitness who has seen Simon Darcourt since Dubh Ardrain.

Of course, in keeping with the grudging and glacially slow nature of what little one could stretch a point as to call progress on this case, it isn't being laid out on a plate for them. More a 'catch and cook your own' offer.

'We got an anonymous tip, via email,' Dale explains. 'Well, ostensibly anonymous: the ISP

277

trace shows it was sent from the company the guy's wife works for. Says the guy has seen Darcourt but is reluctant to come forward because he's scared. Clearly, she wants him to speak up but doesn't want him to know it was her who put him on the spot. We need to tread very lightly on this one, which is why I reckoned a woman's touch . . .'

'Got you. Text me the name and address.'

'On its way.'

THE TRANSFORMED MAN

Well, haven't I just got them all guessing? Ooh, the tension, the suspense, the deliciously tantalising frustration of not knowing. More speculation—not to mention *genuine* excitement—over the result of this poll than the last two General Elections. And what's got them dancing on a string, struggling to contain themselves, like kiddies outside a locked bathroom door after way too much skoosh, is that unlike those elections, right now there's no guarantee they'll ever be told the outcome, and that's just fucking killing them. Wonder how much money I could auction the answer for at the moment, what any of the tabloids, or maybe Sky News, might pay for this particular exclusive.

Truth is, I'm looking at the winner right now. She's lying next to the bag I'm about to fit her body into: one of those big nylon affairs with several pairs of rollers, for taking your golf clubs on holiday. What a prize *she's* bagged, too, if you'll forgive the indulgence of a little pun. The country's never been so interested in her throughout the entirety of her cheap and desperate little life. Wasn't that what she wanted? And not forgetting the runner-up prize, that's a beauty too. Come on, Bully, let's show her what she could have won: a nation's sincerest concern for her welfare. Promises that they'd do anything and everything in their power to save her. Oh, the fucking humanity. Stop it, I've got something in my eye.

So who gets what? Who was the winner? Oh tell us, Simon, please, please, please just tell us.

Fuck's sake, what does it matter? Flip a coin: that's what I did. Jesus, did these cunts really think I sat totting up what was in all those papers, or fast-forwarding hours of Sky-Plus recorded telly coverage in order to gauge who was getting the most play? Did I fuck. And if I was interested, I wouldn't have had to bother, given the number of estimated (if contradictory) running totals that were being kept by the media themselves.

Utterly fucking pitiful, and nothing more than a procrastinatory distraction to focus upon in the absence of any apparent purpose to the whole thing. But now I've closed that option, and it's time for them to really embarrass themselves by filling their pages and their airtime with stabs in the dark as to what it all means, both for me personally and—God spare us—for society as a whole.

In short, what they are looking for is a motive, a grand scheme to explain why I have visited this upon them. It was easier back in the days when they could just use labels like 'sick', 'monster' and 'evil' (all the more convincing reversed out of black in full caps and a massive font size), but in the information age, even the average *Sun* reader knows emotive labels still don't actually tell you anything.

What is the bigger picture here, they demand to know. Why is he doing it?

Happily, I can assure them that there *is* a bigger picture and a very grand scheme; grander than they can anticipate, anyway. But as for motive? Tsk, tsk, tsk. I'm very disappointed that they're still struggling with that one, when the answer is so obvious. A little longer and I really will have to spell it out.

My motive is the oldest and simplest one known to humankind. It is the same as motivates every man who is mindful of his responsibilities and aware—rather acutely aware, as it happens—of his own mortality. Like everybody else, I simply want to make the world a better place before I leave it, and I want to make provision for what will remain once I'm gone. After all, I don't have only myself to look out for, and I won't be around to do that for much longer.

Surprised? Why should you be? Allow that life can change a man, even a man such as me, and especially a life such as mine. I don't have my hand out here, by the way: I'm not saying it's changed me into something that normal society would wish to include, but I am no longer the man I once was. I don't even look like him. I haven't been the same man since Dubh Ardrain, and not quite for the reasons you might anticipate. It wasn't some Damascene epiphany either: more of a slow burn, a cumulative result of a number of factors gradually coalescing. Nobody is ever fully aware that they're changing while that change is under way. At best you only notice it once it is a *fait accompli*, and even then it usually comes long after the fact, but I came as close as you can probably get to realising that some process *was* under way: some chain of events to which I could respond but which I could neither control nor affect. The only way to describe it is go back there, and to look at how the world appeared through my eyes at that time.

It was summer of 2002: almost a year had passed since my personal Waterloo, but there were ways in which I was still reeling from it. Physically I was back to full fitness, but mentally I was suspecting certain damages were done from which I would never recover.

Put more eloquently, indeed put as eloquently as any wright of the language has thus far proved possible, I had of late, but wherefore I knew not, lost all my mirth.

There were a number of evident and plausible reasons for this.

When I looked in the mirror, I no longer saw the face my mind expected. This wasn't any pretentious and self-pitying psychobabble, I should stress: I paid a maxillofacial surgeon a shitload of money to *ensure* that I no longer saw the face my mind expected; and that more importantly, I no longer saw the face so very many people in so very many countries would dearly love to get into with a floorsander and a bucket of hydrogen chlorate. I didn't look radically different. With the bruising and swelling gone, the features and proportions were halfway familiar, but the lines and contours seemed softened; blurred almost, so that I resembled what could best be described as a Japanese *anime* version of myself. I looked different enough, though, let me tell you. Walk into the bathroom for a pish in the middle of the night, catch a sideways glimpse of *that* in the mirror and you're swapping your cock for a Glock, if you haven't already jumped backwards into the bath. When it comes to undermining your sense of identity, having your own coupon replaced would do it every time.

But that wasn't it.

I was sitting, as I did most days, outside a bar overlooking the beach. Resting on my table were a beer and a book, though again, like most days, I was too distracted by what was going on in front of me to read it. This day, however, unlike most days, my attention was enticed by something other than the cornucopia of sun-worshipping females parading between the terrace and the sea. These last had captured my eye but not my mind, a pleasant and picturesque backdrop to my necessarily ugly reflections; soothingly incongruous, disposably irrelevant. I didn't go there for the reasons everyone else does. I wasn't on holiday and I wasn't looking for parties, romance or even just sex. I needed to be somewhere fluidly transitory, where people come and go and the locals don't bother learning your face because you'll be history in a fortnight. I also needed to be in a place where wearing sunglasses from dawn to dusk did not look suspicious or even affected.

That was my life: for the time being, and for as long as needed be. Forever, if necessary, as long as I lived modestly, which was in any case a necessity for purposes of discretion. Not so bad, you might think, and I was always prepared for the possibility, every day, on every job. However, I never considered it an eventuality, and certainly not an aspiration. I didn't do the things I did merely so that I could afford to loll in the sand for the rest of my days. You lie in the dirt long enough later on; what good's a head start? I did what I did because it electrified me every moment I was awake, and I did what I did because I was the best in the whole wide fucking world.

I had killed more people than I could accurately count: four hundred at a rough guess. I had brought down aeroplanes, sunk cruise liners, even trashed a fully armed military base. I had the police of half the planet trailing in my wake, presidential sphincters tweeting at the mention of my name. So all things considered, it was not my idea of the good life to be just another nobody vegetating there in the sun, my back resting against a pile of cash, like some Cro-Magnon Cockney gangster. That was my new life and it represented, to say the least, a bit of a comedown.

But that wasn't it.

I was watching a man and a young child, little more than a toddler, play on the sand with a lightweight football. They were using a pushchair and an icebox as goalposts, the child running up unsteadily to take a shot as his father crouched in the centre of the target. The child connected, giving it a clumsy but firm toe-ender. The ball wobbled in the air as it flew, lending plausibility to the man's transparent attempt to appear wrong-footed. He collapsed to the ground, flailing an arm as the ball bounced past him. The child jumped, hands raised. The man thumped the sand, feigning the anguish of defeat.

Sitting in the bar, there was no need to fake it. I had lost all that I had lost because I had been—and I was—defeated. In my line of work, you didn't lick your wounds then return to the fray with renewed determination; not when defeat meant that the world knew your true name and your true visage. In defeat you may live, but not to fight another day. You may live to become faceless, to drink beer in a reassuringly crowded holiday resort, and to contemplate the person you will never again be.

The sting of humiliation fades with time, but the loss remains, joined soon by a colder, more sober process of recrimination. The apportioning of blame—so often an opportunity for deflection and plain old denial—was simplified for me in that none of my comrades survived. That only left myself and the man who laid me low: Larry the Little Drummer Boy. As my adversary he was responsible for my defeat but not for my failure, so, much as I detested him and much as I resented what he had done to me, I knew it would be foolish and unhelpful to focus my anger upon him. Sitting in that bar, on that beach, I had come to understand that there were occasions when the pursuit of vengeance was simply undignified. Yes, I could kill him, but what would that prove? That I was the bigger man? That he was wrong to cross me? No. Because the truth was that I crossed him and, no two ways about it, I got my arse felt. Twelve of us, professionals, armed and prepared, against . . . well, best not dwell upon the details. When you lose despite such odds in your favour, you have to accept that you humiliated yourself. Seeking vengeance only compounds it. Let's be honest, there is no retribution for a humiliation of that magnitude. Nor is there possible reparation to those who were counting on you.

I failed. Me. *I* was gubbed. *I* was humiliated. My reputation was effectively erased, as surely as I knew my identity would have to be also. Even my memories were all but stolen: when I looked back upon the things I had done, I could no longer view my victories as anything other than a prelude to my ultimate defeat.

But believe me, that was still not it.

My nemesis, my embarrassingly improbable nemesis, did all of this to me. He destroyed my great scheme, wiped out my crew and even cast me down to what he reasonably assumed would be my death. However, my greatest wound, the strike that had me reeling ever since, he delivered with mere words.

There were four teenage males running along the sand, lanky and awkward, suffering that phase nature has the decency to hide inside a cocoon in other species. They bellowed guttural laughs as they bore down obliviously upon the man and his child. The child instinctively moved closer to the man as the group approached, seeking security, protection. The man smiled down, offering reassurance with a ruffling of the hair, but his eyes remained vigilantly upon the teenagers, positioning himself to deflect any accidental contact.

The man was about my age, I estimated. He looked younger when he smiled, but his true years were revealed as his face sharpened in ready defence. The child resembled him facially; I could see that even from the bar. Even if he didn't, there'd be no questioning the relationship: the man was alertly attendant upon the child, but the clincher was that the child looked up at him as though the world was his to command.

'How does it feel to know you'll never see your son grow up?' I asked him, Raymond Ash, my improbable nemesis, when I thought I still held the power of life and death.

'You tell me.'

Now, I analysed and deconstructed this little exchange many, many times, in order to exhaust every avenue of interpretation, but even as I did so I knew I was merely trying to find an escape clause in the small print. I had a gun pointed at his head, so he had to say something to buy himself some time, surely? Granted, but it was still a hell of a thing to just pull out of your arse at zero notice. And as he said it, there was a cold sincerity about him, a conviction that couldn't be entirely accounted for by mere anger or hatred. Under the circumstances you could hardly have described it as smug, but it was definitely the look of one who knew he had something on me; he wasn't only telling me I had a son I'd never met—he was telling me he had.

I had a son.

Through simple deduction and arithmetic I knew who by, I knew where and I knew when. But I did not know him, not even his name, and there were insurmountable reasons why he would never—let's face it, must never—know me.

So how d'you like them apples?

I didn't want a child. Like that needs to be said. Hard to imagine fitting much in the way of family life around a busy schedule of assassination and wholesale slaughter. But discovering, knowing he was out there ... oh Ray, Ray, you really stuck it to me, didn't you? He was loose in my head, toddling around, opening lids and doors and closets, and I seriously didn't want him to see what was inside any of them.

That, in case you haven't guessed, *was* it. He was in there, and he was running the show, whether I liked it or not.

I had in the past transformed myself, or at least attempted to do so: cast off old trappings and

287

emerged as what I imagined to be something new. But whether it was swapping my Queen records for Bauhaus or my Stratocaster for Semtex, the person inside never changed. Larry the Little Drummer Boy called that right. This, though, this truly felt like a transformation by some ancient power far beyond my control. This, as Freddie and the boys put it, was a kind of magic.

I was forced to see the world through my son's eyes, as though new to me, my perspective involuntarily transferred. It was an attempt, if not to feel him, then at least to feel what it was like to *be* him. Then I saw it once more through my own, and I felt a dread darker than anyone else could know; well, maybe not anyone else, but we'd be talking about a very short list. I felt a dread because I knew what kind of men were sharing this world with my son.

Evil men. Men like me.

I was feeling a new, alien emotion. Maybe it wasn't alien, maybe it wasn't even new. Maybe it was an instinct that had always been there, dormant, only recently activated. I felt a drive to protect, an anxious concern for this child of mine whose face I had never seen and whose name I did not even know.

I sensed his vulnerability. That was something for which I'd always had a facility: I sought out the undefended, probed for weakness, then gleefully exploited it. This vulnerability instead put me on edge, compelled me to vigilance. I considered the carnage I had strewn about me, and I was appalled to think of him getting caught up in something like that: among the gormless, faceless lemmings who just couldn't help but find themselves in the wrong place at the wrong time. He was worth more than that, more than them. Far more. He was flesh of my flesh, my son, and the thought of someone harming him did not merely worry me—it offended me.

Deeply.

It was for that reason, therefore, as I sat there watching another father tend and protect his offspring, that I felt unexpectedly inclined to act upon certain information which had, that day, become pressingly pertinent, but to which I would previously have been utterly indifferent.

A four-year-old English boy had gone missing from one of the big villas on the other side of the Old Port. It was all over the island that morning; everybody knew about it. That was why the father on the beach, like every other parent there, was staying that bit nearer to his precious child, watching that bit closer, grateful he has not been punished for lacking the same vigilance twenty-four hours ago. The kid's face was on the front of the local paper, and the police, plus dozens of volunteers, were combing the area. Divers would be brought in too, inevitably, but only if the parents were smart enough not to tell the cops about the ransom demand they were about to receive.

I knew this because I knew who had done it.

All life passed that bar, there in my sun-kissed purgatory, and it didn't have to be wearing a bikini to catch my eye. It therefore failed to escape my attention on either of the occasions that Risto Balban and his moron brother, Miko, sauntered conspicuously along the boardwalk, having evidently travelled to the resort in the past month with Club Thug. Risto used to be a big noise in the Kolichni separatist rebel movement, which employed his kidnapping and extortion skills to political and fundraising ends. Some of those funds ended up in my pocket for services rendered, which is why I knew his face. But this was before his political convictions waned in the face of his realisation that he could get up to the same hi-jinks independently, without having to hand over the resultant cash to any pompous ideologues in balaclavas.

It was common knowledge within certain less-than-exalted circles that he and Miko had been busying themselves around southern Europe ever since. They targeted the most upmarket holiday residences (not much ransom to be got out of the *Sun*-reading classes) and went for kids of four years and under because they didn't tend to be much cop when it came to giving the police descriptions. That was the ones whose parents kept it shut and paid up in time, of course. But despite their industry, you wouldn't have heard about any of this, because the authorities in tourism-driven economies could bring rather a lot of pressure on the local plods regarding their after-the-fact interpretation of such events. Who's going to go down to Lunn Poly and book up for 'that resort where the wean got kidnapped last summer'? So the local kiddy-fiddler gets fitted up for murder, the 'isolated incident' is solved, and the Balbans move on. They'd worked Greece, Turkey, the Black Sea, moving from island to island, coast to coast.

And now they were here. Risto, the brains of the operation: lithe, sharp-featured, canny, paranoid.

Miko: tubby, thick, obedient and loyal, as proven by the metal holding his legs together since a mutually unsatisfactory interrogation at the hands of the Russian military. I knew what they were here for, I knew what they would do, and at the time it didn't seem to be of any import. I had always considered their activities vulgar, but none of my affair; indeed my principal concern when I saw them was whether my surgery would pass the eyeball test. However, now that they had actually carried out their work, I found myself experiencing an unaccustomed outrage that I knew I would not be able to contain.

*   *   *

I sat in my car and watched the villa, easily identified by the police cars at the gate. I waited. I would give it an hour, I decided. This particular exercise was to ensure that my good deed went unpunished, but there were other ways of doing that if an opportunity failed to present itself that morning. After twenty minutes, however, it did. I saw the father walking out of the driveway with a pushchair, occupied by a tiny infant. He stopped briefly to exchange words with one of the cops, gesticulating towards the buggy. He was telling them he was just taking the little one for a walk, but his other hand looked like it was the only thing holding his head on, so I knew differently.

I got out of the car and paced myself to catch up with him out of sight of the police. He stopped at a bench overlooking the water and sat down, offering the infant a bottle of water. The infant smiled at him and he smiled back, trying to hide how he was really feeling. I didn't know whether the baby was buying it, but I certainly wasn't.

I sat down next to him and spoke facing out to sea. I wore large shades and a hat. Not an impenetrable disguise, but it concealed enough to blur any future picture.

'I know what you're thinking,' I told him. 'And the answer is no. Don't tell the cops.'

'What did you say?'

'Ransom demand. Phone call. Voice disguiser, right?'

He stared at me, standing up. I remained seated, looking out to sea.

'Who are you? How do you know this?'

'If they even think you've told the cops, they'll kill the boy and move on. To them, it's not worth the risk; they can start again elsewhere tomorrow. You can't.'

'Listen, tell me who the hell . . .'

I turned to look at him.

'I'm going to bring him back.'

'You're going to . . .' Confusion and indignation gave way to desperate hope. 'How?'

'That's my concern.'

'I don't understand. Who are you?'

'I'm in a position to help, that's all. I can't say any more.'

'And what's in it for you? What do you want, money? How do I know you're not part of this?'

'I'm not. And I don't want money. But I do need two things from you.'

'What?'

'First, that you follow my advice and don't tell the cops.'

'Christ, I haven't even told my wife yet.'

'And second, that when I return your son, you forget I ever existed.'

'If you return my son, I'll remember you for the rest of my life.'

'No doubt, but you don't need to tell anyone, do you?'

*   *   *

In tracking kidnappers, knowing whodunit is far less crucial than knowing where they are, but if you know enough about the who, ascertaining the where is a straightforward—if sometimes necessarily messy—process. In this case, it was enough that I knew that the Balban brothers, not having much of a portfolio, preferred to invest their disposable income in concerns yielding a more immediate dividend: to wit, sex and charlie; or more accurately, hookers and charlie, given that the concept of Risto or Miko getting laid without paying for it was known in most cultures as rape. I knew also that they were unlikely to be indulging in the former vice while they had a houseguest, so consumption of the latter was bound to increase by way of compensation.

I called several local suppliers, with whom I had, shall we say, a rapport, and ran message-boy Miko's description past them. Arturo at the casino (where he was by far the busiest dealer), confirmed a portly source of much recent custom. I made a business proposition and we arranged to meet within the hour. Upon my instruction, Arturo rang the mobile number Miko gave him and offered him the drugs equivalent of a fire-sale: he told him he was having to get off the island in a hurry, and

293

as he couldn't take his stash with him, the whole lot was available at a knockdown price, for a limited time only. Hurry, hurry, hurry. Miko went for it with all the restrained dignity of a piranha, agreeing to a meet early that evening. I paid Arturo the shortfall in the price arranged with Miko, plus a little more for his trouble, then headed for a tool-hire outlet in the industrial area of town.

*   *   *

I watched Miko emerge from a BMW and approach the casino, walking as ever like a gorilla bursting for a shite. He was dressed in an evening suit in an attempt to look inconspicuously respectable, but given that primate gait and his mangled features, the jacket and tie in between looked sufficiently incongruous that he might as well be wearing a t-shirt stating 'In-bred gangster trash'. I watched him enter the casino, then tried his car door and found it unsurprisingly open. It's a common conceit among these criminal also-rans that people—especially their fellow crooks—are somehow aware of how baaad they are, and would therefore never dare fuck with their person or their property. I lay down in the rear footwell and prepared to disabuse Miko of both presumptions.

He returned inside ten minutes, giggling like a kid who's raided the candy shop. I waited for him to place his prize in the glove compartment and reach for the ignition, before spiking him in the neck with a hypodermic full of thiopentone. He reacted reflexively with a slap, thinking the jab was an insect bite, by which time the agent was already in his blood and my silencer in his face.

'Sweet dreams,' I told him.

*   *   *

Miko awoke to find himself strapped securely to a steel table in a low-ceilinged, windowless room. We were in the cellar of my villa, in the hills overlooking the port, but Miko didn't know this. Nor did he know that he was about to become nostalgic for the hospitality of those Russian soldiers.

He came round slowly, groggily at first, but sharpened up very quickly as he took in his surroundings and realised the circumstances. I remained behind him at the head end of the table. If he strained his neck muscles he could see me, but for the moment he was scanning the cold stone walls, bare apart from cobwebs and the ancient, dustbound workbench to his right. On it sat a rusted but serviceable vice, a power drill, an electric paint stripper, a hacksaw, a boombox and a feather duster. His breathing accelerated and his arms tested the restraints. I guessed it wasn't the duster that spooked him, but some people do have awfully ticklish feet. You never can tell.

Predictably, he asked who I was, trying it in French, English and Spanish.

'Let's save time and just lay our cards down on the table, Miko,' I said. 'I want to know where Risto is holding the boy.'

'How do you know my name?'

'I know a lot more than that. Most relevantly, I know that I would be insulting you to assume you'd give up your brother without first undergoing some

295

quite unnecessary prolonged and excruciating pain. Equally, you would be insulting me if you thought I'd believe anything you said before that, so allow me to treat us both with all due respect.'

His eyes flitted to the workbench again and he swallowed, a look of determination fixing upon his features. Then I wheeled the gas tanks from behind the table, into his line of sight, which is when he started to squirm and whimper. I flipped down the visor on my protective mask and pressed Play on the boombox. The sound insulation in the basement was fine, but I find screaming to be very disconcerting while I'm trying to concentrate. I fired up the torch as the music started. It was Neil Young and Crazy Horse live: not strictly my cup of tea, but I thought it appropriate, though I doubt Barry Sheen there got the gag. The album was called *Weld*.

It fairly grew on me, I must admit, though not as much as what grew on Miko.

I stood back when I was done, waiting for Miko's hysterics to exhaust themselves. There was a strong smell of burnt meat filling the room. I'd read medical staff distastefully describing the odour of charred flesh, and I imagined it must be pretty rank by the time you got it down to Casualty, but right then it wasn't a kick in the arse off barbecued chicken.

The steel in the lower half of Miko's left leg was now fused with my table. It wasn't a very professional-looking job, and the table would never be the same again, but Miko was going to have a bugger of a time going anywhere without it.

'So, Miko. Bearing in mind that you have a lot more metal in those legs of yours, do you feel like telling me where Risto is holding the boy?'

He was hyperventilating wheezily, but I could tell he was summoning the breath to speak. I leaned closer and he told me the name of a villa outside Fornel, about forty minutes away on the road to the airport. I got him to repeat this, then reached for a drawer under the table.

'Time for me to leave you, then,' I said. 'But before I do, one more thing. Don't take it personally, but it struck me that you just might be lying, you know, maybe to buy yourself time so that you could forge an escape. No pun intended.'

I produced two syringes and lay them on the workbench where he could see them: one containing clear fluid, the other a pale blue liquid. I picked up the clear one and injected its contents into a bulging vein in his forearm.

'This is dihydromertile silicate,' I told him. 'It's slow-acting, so you won't feel anything for a while, but it will stop your heart and your lungs completely in about two hours. I know you weren't paying much attention on the trip here, so I should let you know you're in an abandoned farmhouse in pretty much the middle of nowhere, and I'm afraid there's little chance of someone stumbling across you and coming to the rescue inside the time you've got left. However, on the upside, the blue syringe contains a neutralising antidote: monohydrate dosamide, and I'm going to leave it here. This is the deal: I go to Risto and, to avoid an unseemly squabble, I offer the location of this farmhouse in exchange for the kid. He saves you, I save the boy, and we're all happy. Unless, of course, you're lying, in which case I administer the antidote and we listen to Neil Young all night

297

long.'

Miko closed his eyes, steadied his breathing, then in a broken whisper told me where Risto *really* was.

*   *   *

The villa was set, rather picturesquely, in a sprawling vineyard, with high hedges thoroughly obscuring the building and its gardens from the road. As I turned into the vine-flanked avenue, the headlights of Miko's Beamie flashed across a curtained window. Thus pre-warned, Risto emerged impatiently from the front door just as I pulled up, silenced Glock out of sight beneath my open window. I shot him in both legs before he could speak, then stepped out of the car and knelt on top of his writhing body, patting him down for weapons. I found a nine-mil and a stiletto.

'Were you expecting someone else?' I said quietly to him. 'Anxious times when family goes missing, aren't they? Where's the boy?'

He looked up at me, his eyes barely able to focus for the pain. 'Fuck you,' he managed to splutter.

'Don't worry, I'll find him myself. Can't expect you to help me with a bullet in your balls, can I?'

I let the remark register for a second. He pulled a single key from his trouser pocket and let it drop on the flagstones.

'Thank you.'

I found the boy tied to a bed, gagged and urine-soaked. He flinched as I approached, and I remembered I was carrying the Glock.

'It's okay. I've come to take you home.'

I removed his bonds. He looked uninjured but remained terrified. When he spoke, his voice was lower and more croaky than I expected, due to dehydration.

'What about the bad men?' he asked.

'They're very, very sorry. They won't be doing it again.'

'I want my mummy.'

I lifted him up with my left arm, raising the gun again with my right.

'I'm sure they do too,' I said.

I carried him with his face rested on my shoulder, and told him to keep his eyes closed until we were in the car. I fastened his seatbelt and turned on the engine, then climbed back out again.

'I'll just be a second. I'm going to get you a drink, okay?'

The boy nodded, still trembling.

I dragged Risto inside the house and out of sight. With the lights on in there, he was able to get a better look at me. The surgery was still making me hard to place, but I obliged him with a lingering stare into his eyes until he recognised mine.

'My God. You're . . . you're . . .'

'Not any more. I'm just a concerned parent.'

I put four bullets in his brain then headed for the fridge to grab a Coke for the kid.

*   *   *

The boy obediently kept his head down as we neared the agreed rendezvous. I donned the shades and cap again as I drove past the bench where we spoke that morning, then doubled back, checking the dimly lit side-streets for any concealed cop cars ready to swoop in. It looked as though the father had been true to his word. He was sitting there, looking expectantly at the Beamie, as he would have done at every other car that had passed since he arrived.

He sprang to his feet the moment his eyes met mine. I stopped the car but didn't get out, merely reached back and undid the kid's seatbelt. The father opened the door and hugged his son, both of them crying. I looked away.

'There must be something I can do for you,' he said.

'There is. I told you.'

*   *   *

When I got back to the villa, the first rays of sun were still loitering with intent behind the hills, the air pleasantly crisp before the heat started to build once more. I dropped Miko's bargain charlie in my safe then went down to the cellar to find the man himself. The scene did not disappoint. He was dead, face-down on the floor, an empty syringe lying discarded beside him. As I anticipated, he had freed himself from his restraints through brute-strength and desperation, but with his leg welded to the table, was only able to reach the hacksaw, not the hypodermic. He had proceeded, therefore, to amputate his own foot, before injecting himself with weed-killer in a misinformed attempt to neutralise the harmless saline solution I gave him earlier.

It's uncool to laugh at your own jokes, but I couldn't help it. Maybe I was getting whimsical now that I was technically a generation older; and I know I wasn't around for the punchline, but you have to admit it was a belter.

I went back upstairs, grabbed myself a cold one and sat outside to watch the sunrise. It felt like a new beginning . . . but just for a moment.

What I couldn't get out of my head was how I felt towards that child's father: even when I saw him in his earlier anguished state, I realised that I envied him. I would never, could never have the life he had, but that wasn't it, Christ, no. He was still just another Suburban Sad Cunt, but to my surprise I suddenly understood the SSCs' secret. All that guy wanted was the kid back, and not because losing the kid had changed anything. All he wanted the day before the kid went missing was to be with the wife and weans. I didn't envy what he had, or what he wanted. I envied him *that* he wanted it. I envied him that it was *all* he wanted.

I envied him that it was enough.

THE OLDEST MOTIVE

His name is Neil Baker. He agrees to meet Angelique at a motorway hotel on the M25, a place where he frequently convenes with business clients in one of its smaller conference rooms. It takes some persuading, and a lot of reassurance, but she manages to secure his assent to work with their artists, though again, this will take place in another hotel conference room, to which nobody will arrive in a marked police car. At last, they'll have an image, but they'll give no indication as to its source. It will be a charcoal sketch based on five-year-old memories of a partially disguised man only glimpsed for two short moments during which the guy's brain was in meltdown, but it is that kind of case. The very fact that they will have it at all appears to be the sole tangible return on the only previous clue they've been able to offer the public.

'It was the voice,' Baker says. 'As soon as I heard it on TV, I went rigid. I can remember everything he said to me, and I hear it replaying in my head any time I think about what happened. I remember his voice far more vividly than I remember what he looked like, maybe because after our first encounter, I kept repeating his words in my mind, kept going over and over what he had said, like you analyse every fibre of anything that might offer hope.'

Angelique doesn't find it necessary to ask why he hadn't come forward.

'I was terrified,' he volunteers. 'Christ, once I realised who Josh's angel of mercy had been, I was physically sick. I spent years wondering who this scary character was, concocting b-movie notions of rogue vigilante cops and gold-hearted villains policing their own. I took it as reassurance that there existed in the darkest hearts some innate sense of decency. Then I heard that recording, and I've barely slept since. I don't understand why he took it upon himself to help us, but I know for bloody sure I don't want him taking me to task for ingratitude.'

He doesn't ask who tipped them off. Doesn't need to, she guesses.

Angelique, in accordance with the man's request, walks away and heads for the car park, allowing him to exit alone later, once she is gone. She thinks he's going a wee bit too belt-and-braces with the anonymity measures, but she can relate to why: if you've had a one-on-one with Simon Darcourt, you generally wouldn't be in a hurry for a reprise. Not unless your parents' lives depended on it, anyway.

*   *   *

You had to give this to Darcourt, he was always good for a mindfuck, and this one, for Angelique's money, topped the lot. Going out of his way to help out a complete stranger was, in its own way, the most perverse thing he had ever done. Darcourt never did anything without a very strong personal motivation, even if that motivation was electronically transferred into a bank account in Lichtenstein. This guy had neither given nor offered any reward, so what had driven him to rescue a kidnapped child? Insane as it sounded,

303

Baker had perhaps stumbled on to a truth with his previous speculation about an innate sense of decency in the darkest hearts. There was no heart darker than Darcourt's, and no sense of a value system that anyone else would recognise as decency. But as Ray had put it, 'he was one of the most self-righteously vindictive people on the planet': thus, he did have his own skewed, inconsistent, self-serving and hypocritical but very rigidly enforced morality.

Something about the kidnapping had offended that morality. Darcourt's conscience had previously been untroubled by the welfare of the many young children who had died in the atrocities he had engineered, so something about this was different. Either that or something about *him* was different.

Ray.

Climbing into the car to escape the blustery drizzle, she reaches for her mobile and listens to it ring as the rain warps her view of the passing traffic. Ray takes a little while to answer; he's probably in class.

He confirms as much when finally he picks up, asks if it can wait. It can't.

'At Dubh Ardrain, did you tell Darcourt he had a son?' Angelique asks.

Ray pauses. 'Yeah,' he confirms grimly, clearly appreciating the altered significance of this, given that Simon is still alive. 'He was about to shoot me,' he explains. 'He was milking the moment, asked me how it felt to know I'd never see my son grow up. I just said: "You tell me." I was trying to mess with his head, to distract him, buy myself some more time. Christ, I hadn't thought about it

304

till now. Somebody's got to warn Alison and Connor.'

Angelique assures him that they will be protected. In a way, they always have been: there were a few 'I slept with terrorist monster' chequebook kiss-and-tell pieces after Dubh Ardrain, which had largely satisfied the press's appetites regarding the Black Spirit's personal background, but in case these didn't, there was an injunction in place to prevent Alison and Connor McRae from ever being named in connection to Darcourt.

She terminates the call quickly but politely, then flicks through her contacts list, deciding who best to phone regarding the immediate surveillance and protection of the McRaes, in case Darcourt comes looking for them.

That's when it strikes her that he already has.

'I can't see Simon tuning in via satellite from some mainland European bolthole just so he can stoke up his rage,' Ray had argued, back when he was still trying to convince himself that the new killer wasn't his old flatmate. 'Can you picture him getting the *Sun* or the *Daily Mail* delivered to his underground lair or wherever he'd have been hiding?'

Now she could see the consequences of that same logic, once it was inverted. He wasn't in some European bolthole; not all the time, anyway. He had spent time—maybe a lot of time—back on his native soil, in order to get close to his son: disguised, unsuspected, assumed dead, watching secretly from a discreet distance, maybe outside school, the swing-park, the sidelines of a football pitch. Near as he dared, perhaps, but no closer than risk allowed, which was why, though she would still call in the warning, she understood Connor and Alison were in no danger. If he was going to snatch the kid, he'd have done it by now, before anyone discovered he was still alive. He knew he could only watch Connor's life like it was on the other side of glass. Not only could he not risk making any kind of direct contact with the boy, if there was any residual humanity in him, he would understand that he couldn't do anything that increased the likelihood of his son finding out who and what his father really was.

In Darcourt's distorted mind, it was thus possible to imagine him finding a surrogacy in acts of vigilantism such as the Baker child's rescue.

You only had to peruse the letters pages of any newspaper, or these days, any internet forum, to discover the extent to which the average bampot considered parenthood both a justification of and a sanctification for their own self-righteousness. What would it do to 'one of the most self-righteously vindictive people on the planet'? What other deeds might he consider proxies for the actions of a concerned father unable to directly shape and protect his offspring's development? Every sad dad tries to make up for his own broken dreams by living vicariously through his children. So wouldn't he want a fairer world, in which the truly talented, like him, got rewarded with riches and acclaim, while the mere attention-seeking mediocrities had their wings melted for even daring to fly close to the sun?

This new perspective potentially altered the time-line too. The big question they'd been asking since he made his theatrical reappearance was why now? It had only been lately that he had revealed his survival and reclaimed his name, but what if they were wrong to assume that this was the beginning? Maybe it was merely the next phase. The Baker kidnap had been more than five years ago. What else might he have done in that time by way of moulding a better world for future generations, in between visits back to the mother country to steal a glimpse of his son and heir?

And with this question, another, older, unsolved mystery may suddenly have a new solution. Angelique remembers a case Dougnac was called to investigate, in Lombardy, four years back: one the public got to know nothing about. The official story, as reported briefly in the press, was that half a dozen 'aviation industry' executives had died on a corporate junket when a yacht went down in the Med off Genoa. There was no such incident: the 'maritime tragedy' explanation was concocted at the behest of high-level influence, the shipwreck scenario chosen because it accommodated the absence of any bodies to fill the coffins at the six respective funerals. In truth, for 'aviation industry', read 'arms trade', and the reason their coffins were buried empty was that, short of DNA sampling, there was no way of knowing which particular puddles of viscera belonged in each casket.

Dougnac was brought in to explore any possible terrorist angle, but the absence of any claim of responsibility or discernible ideological motive had him quickly rejecting the idea. The execs all worked for arms manufacturers, and it was his opinion that the incident represented the sharp end of industrial relations. These firms were often in bed with some extremely dangerous people, and the political influence brought to bear in order to conceal the true nature of the murders was proof of how far—and how high—they were prepared to go to prevent their dirty laundry being aired in public. Dougnac's suspicion was that the murders had been ordered and orchestrated by none other than Marius Roth, for business and power-broking reasons that only the highest-placed within the European arms trade would ever know.

Angelique had reckoned her boss's logic was sound, but maintained private reservations regarding his conclusion, on the grounds that Roth had always been the madman in Dougnac's attic: his *bête noire* or, perhaps it would be more revealing to say, his white whale.

Marius Roth was a quite squalidly wealthy arms industry maven, mogul and manipulator. He exerted overt and covert influence in the boardrooms of countless 'defence' companies, brokered deals involving arms firms and governments around the globe, sanction-evasion and embargo loopholes a speciality, and couldn't have had more politicians in his pocket if they were miniature-sized and made by Playmobil. All of which, of course, was merely his public, 'respectable' face. It was what remained shady and occluded (and the ways in which the shade was cast) that led to such mystery and rumour, not to mention myth-making, as proved fascinating to the point of obsession for Dougnac. Roth's greatest value to the arms trade was said to be his equally influential manoeuvrings in ensuring that consumption of its products remained high. He was believed to be just as senior a power-broker, fixer and facilitator in the world of the freedom fighter, guerrilla, rebel, insurgent or, if you were to be so crude, terrorist, as in the world of the executive and the politician.

However, he was also a world-class master of deniability, so nothing could even remotely be proved. Intel on his more occult connections was frustratingly thin, with sometimes only the conspicuous *absence* of evidence constituting the only clue. A marked tendency for suspects believed to have had dealings with Roth to die before the authorities could talk to them was one such conspicuous absence. Many died in custody, some disappeared *from* custody, never to be seen again, and some, most alarmingly, killed themselves, usually after going to extremely desperate lengths to avoid being taken alive. Dougnac believed that they did not merely fear the consequences of giving up information, but that if they were captured at all, they were immediately tainted and thus condemned in the eyes of the people they were afraid of. No amount of swearing you told the cops nothing was going to save you. What it wouldn't save you *from* was evidently the stuff of nightmares even to hardened men of violence.

Throwing another layer of gauze over Roth's blurred picture was the fact that, even were the authorities to keep a suspect alive long enough to talk, if they were high enough up the food chain to have had direct dealings with the man himself, they wouldn't necessarily *know* that they had. It was said that while Roth necessarily maintained a recognisable (albeit low-profile) public identity, in his darker dealings, he was a shape-shifting wraith who never gave the same name to two people, and who altered his physical appearance to further distinguish these multiple personae. Dougnac said they had some extreme-distance zoom shots showing him bald, hunched and pot-bellied; others slim, upright and with a full head of hair. Dougnac reckoned he used not just wigs, but various prostheses too, meaning that no two contacts, even if they were to be so suicidally indiscreet, could compare notes and conclude they had been dealing with the same man. These photos had been taken while the subject was aboard a vast luxury yacht, the ownership and registration of which were needles in a legal haystack of fronts, pseudonyms and shell-companies. Furthermore, it had never been satisfactorily established that the man in them was even the same individual—disguised or not—as turned up at board meetings and parliaments. To put the tin lid on it, Dougnac entertained the truly brain-twisting possibility that his 'respectable' public identity could be just another fabricated cypher. 'There may not even *be* a Marius Roth,' as he perplexingly put it.

A damaging symptom of this frustrating elusiveness was that Dougnac could fall into the trap of imagining Roth's hand behind anything and everything, especially in the absence of an alternative explanation. Thus he had been inclined to believe his personal bogeyman was responsible for the Lombardy incident, particularly as there was no apparent motive or beneficiary: two factors that Dougnac expected would remain concealed but, in certain elite circles, known and very clearly understood.

But what if the 'aviation industry' killings had been Darcourt: like the murderer attacking the paedophile, his cracked moral compass still functioning enough to point him towards a target that might paint the perpetrator in a better light than the victims?

The implicit self-righteousness of it was certainly in keeping with Darcourt's new MO, so there was as much reason to pin it on him as there was Roth. Roth's tentacles spread far and wide, but that didn't make him responsible for half as much as Dougnac tended to imagine. Nor was he quite as untouchable as Dougnac's reverent obsession would suggest. Roth had, after all, been visited by a *genuine* yacht mystery of his own. A couple of years back, his vessel *Corsair*—this boat quite definitely registered to the named businessman—was discovered to be drifting off the French Riviera, unmanned in as much as the goons who crewed it had all been shot dead.

Embarrassing and compromising as this incident clearly was, it was testament to Roth's mythical stature—and perhaps his people's spin-control— that the rumour which grew legs quickest explained the killings as an internal punishment: a *pour encourager les autres* exercise in retribution for some undisclosed failure. Angelique, who was less inclined towards chasing ghosts, preferred the more plausible theory that his team had been humped: that someone had boarded the yacht and ripped them all a new one, with extremely disappointing consequences for one or several items on Herr Roth's agenda.

Could that also have been Darcourt? she wonders, for the fraction of a second it takes to remember a conclusive argument to the contrary. To wit, Darcourt is a shite-bag. He always specialises in easy targets: the unsuspecting, undefended and unarmed, so taking on a crew of weapon-toting hired muscle was as outside his MO as it was possible to get. Whoever hit the *Corsair* had undertaken an assault—from the water, no less—against a team of trained mercs armed with automatic weapons and in organised communication via short-wave radio. So while it did prove that Roth *was* touchable, it nonetheless put him in an extremely low percentile of vulnerability. He had nothing to fear from Simon Darcourt, anyway, Angelique muses, before halting that thought on the proverbial dime.

She's staring at the windscreen, the drizzle rendering the glass as opaque as the black surface of a storm-churned sea, but what she's seeing has absolute clarity.

Marius Roth dealt in all manner of 'defence-related' commodities, including talent. Given the Black Spirit's 'have Semtex, will travel' status as a terrorist-for-hire, it was impossible to imagine that Roth didn't have a hand—not to mention a finder's fee—in securing Darcourt's services on behalf of various bloodthirsty and cash-rich psychopaths. General Aristide Mopoza, for example, who had contracted Darcourt for Dubh Ardrain, and who had, now she comes to think of it, found himself assassinated shortly after its failure. Assassination was an occupational hazard in the post of military dictator to a country such as Sonzola, but Dougnac had commented on there being a spooky familiarity about Mopoza suddenly not being alive to talk about something that might potentially lead back to Marius Roth. Darcourt, having apparently died at the power station, posed no such danger. But now that he has revealed himself to have survived, and revealed this to as wide an audience as possible, then it would be fair to say that Marius Roth *does* have something to fear from Simon Darcourt. Enough, in fact, for Roth to unleash his dogs in order to get hold of the bastard before his attention-craving atrocities lead the authorities straight to him.

Marius Roth: the kind of man who has 'sources who do not even know they are sources'. Marius Roth: the kind of man whose people could at short notice arrange and carry out the kidnap of her parents and render her their puppet inside the police hunt for Darcourt. Marius Roth: the kind of man whose retribution haunts the nightmares of even the worst of killers.

She feels the fear flood her again, like it did that night inside the taxi. The phrase 'tortured to death' insinuates itself, threatening her control as it mingles with thoughts of what methods might drive hardened criminals suicidal with fear.

She opens the car door, concerned she's about to be sick, and steps out into the rain, bending over. Nothing comes of it, but the blood rushes to her head, relieving the onset of faintness. She holds still a few seconds, then stands up straight, lets the breezy smir blow about her face. It's cooling, immediate. She's focused again.

This changes nothing, not yet anyway. If Roth wants Darcourt that much, then his people will have to keep her parents alive for now. If it gets that far (and she knows she'll be doing very well if it does get that far), it will still be all about the exchange, and the key to that was handed her yesterday by Albert Samuel Fleet.

She lets the cool of the rain play upon her face a

313

little longer, dousing the heat from her cheeks and easing the tightness gripping her chest. Feeling sufficiently composed to commence lying to a mutually trusted colleague, she climbs back into her car and calls Dale, to play her one and only card.

She briefs him on Baker, adding Ray's confirmation that Darcourt *père* knows there is a Darcourt *fils*. It's while Dale is reeling from the concept of this newly concerned father reinventing himself as a psychotic combination of Charles Bronson and Mary Whitehouse that she finally reveals the lead she's been sitting on for days.

'I think I might have a new angle,' she tells him. 'You remember we struck out with the idea of looking into crooked plastic surgeons?'

'Yeah,' he mumbles, suddenly lowering his expectations of what she might have to offer. What upon suggestion seemed a promising avenue of investigation had very quickly turned into a cul-de-sac, terminating against the concrete wall of medical ethics and confidentiality. There were surgeons out there known to have provided a no-questions service to individuals in urgent need of a new appearance, but not only were they unwilling to tell the cops anything, the slimy bastards also knew there was no way you could access their files.

'Well, I just got a heads-up from a contact back in Paris, regarding a maxillofacial surgeon: one Doctor Guillaume Bouviere.'

In fact, the 'heads-up' had consisted of a text accompanying another image of her captive parents. Responding to one of her updates reporting the investigation's lack of progress, it stated simply:

314

Suggest you seek Dr G Bouviere, who recently ceased to be of use to the criminal fraternity.

'Is he a reformed character or something?' Dale asks none-too-hopefully.

'After a fashion. He ran a private surgery near Toulon until three months ago.'

'What happened then?'

'He was murdered in the car park of his clinic. Rather than *bent* surgeons, I decided to put out a filter on dead ones,' she fibs. 'He was beaten and stabbed, believed to have been killed by junkies looking for drugs. Reports say they took his keys and started tearing the clinic apart before somebody raised the alarm and called the *gendarmes.*'

'They get anybody for it?'

'No. The junkies line seems to have stuck as far as the local cops are concerned. However, my contact says Bouviere was known to have done work for the underworld. Bent facial surgeon dies and his place trashed, just before Simon Darcourt makes his grand reappearance. Gotta be worth a sniff.'

'Not if the killer got away with the files, which would be presumably what he was after.'

'Ah, but he didn't. Like I said, the alarm was raised. Bouviere's files are now securely in police storage.'

'But presumably still subject to the usual confidentiality laws,' Dale reminds her.

'Which is why I'll need seventy-two hours, maybe longer, to go over there in person.'

'What difference will that make?'

'Let's just say it gives you deniability if you don't ask me that kind of question.'

Dale sighs, makes out he's weighing up his options, but even in her state of enhanced anxiety, Angelique knows he's buying.

'If you bring back something that helps us find Darcourt,' he says, 'I'll ask no questions at all. That's a promise.'

Angelique stares at the rain and is grateful she's not having this conversation face-to-face. She'll help find Darcourt, sure, but the real purpose of her trip is to bring back something that ensures they don't get to keep him.

'Trust me,' she replies.

PARENTAL ADVISORY: EXPLICIT PURPOSE

I'm hardly the sentimental sort, but I'll admit, freeing that child from his kidnappers, I enjoyed making a difference. I experienced something I hadn't felt in years, and not as strongly since my first kill: that tuppence-ha'penny little schemie gangster Frank Morris, who I murdered to avenge my father. It was the exquisite satisfaction of doing what justice demanded, and doing it at nobody's bidding but my own. After so long working as a killer for hire, I had almost forgotten how liberating, how much more pure it felt to be driven expressly by my own desire, rather than by simply professional commitment. It was like the epiphany a chauffeur must experience, one day taking the high-end Merc on an open country road just because he feels like it, after years of ferrying suits through the perma-congestion of the city. A sudden, invigorating reminder of the power at your control, the possibilities lying open before you.

However, at that same time, it said much for my altered psyche that I viewed such power, such possibilities, as merely a consolation. There could be no overt role for me in my son's life: I understood that. I didn't even, by that stage, know his name, and by the time I learned it, and had laid my eyes upon him, I knew that he should never know mine.

Unfortunately, given that he was growing up among that nation of curtain-twitchers and hypocritical, gossiping nobodies, there was no way

317

of guaranteeing he would be protected from this knowledge. Thus I toyed for a while with how I might ease the burden, should he ever be shouldered with it. I sought to offer him a different perspective upon me: something that he and the tunnel-visioned nodding dogs who handed him down their homogenous system of values could relate to.

Landmines: they were nasty little fuckers, weren't they? If the late Princess Diana, the true-north on the nation's moral compass, had taken a stand against them (albeit in between many and various helpings of cock, though the guardians of decency seemed to have a collective blind spot around that), then who could imagine worse transgressors against our shared values than the scum who were schilling those babies?

And, to be fair, she had a point. It pissed me off that while my lifetime tally remained comfortably below five hundred yet I was unlikely to be invited to any royal garden parties, these devices accounted for an average of seven thousand fatalities per year, and the heads of the firms who made them were feted as captains of industry—even after the Princess had posed in her flak jacket. Plant a couple of them and you get called a murderer. Plant a thousand of them and you get labelled a warlord. Manufacture a million and you get a fucking knighthood—as long as they're just one sideline in a wide enough range of military or 'aviation' products, and your corporate structure is sufficiently byzantine as to conceal precisely who and what you control.

After the Ottawa Agreement, some of the chancers even had the cheek to claim the high ground, with solemn-faced pronouncements to the effect that they wouldn't be making them any more. This was largely to spare the blushes of the politicians they were tight with, as this public delousing exercise allowed them to keep up their mutually beneficial back-scratching without fear of PR fall-out. It was like an instant clean sheet, sidestepping the embarrassing fact that the mines these po-faced pricks had *already* flogged would be killing an average of nineteen people that very day.

They assured an approving public (and thus terminated that public's interest in the issue) that they would no longer be making these terrible devices, which could render land useless for decades and kill or maim unsuspecting civilians years after the conclusion of the war for which they were deployed. And as far as these specific devices went, they were telling the truth. However, they had little real intention of ceasing to profit from landmines, any more than they had any intention of ceasing to profit from armed conflict in general. Where there's a will, there's a way, after all.

I had my fun with a selection of involuntary industry delegates at an arms fair near Milan. Did I say arms fair? How vulgar of me. I meant to say *air show*. That's a fond front for the murder-facilitation-and-logistical-support industry. This air show was much like an Eighties corner shop, where the majority might innocently browse among the widest spectrum of products: from sweeties in jars to Molly Ringwald flicks on Betamax, via emery boards, oven cleaner and pints of milk; but where the inducted cognoscenti know they can also procure Scandinavian hardcore vids if their face is known and they ask the right guy.

The under-the-counter selection at Milan was showcasing a number of new products. The slippery fuckers had been through the Ottawa Agreement with a fine-toothed comb and were already engineering their way around its definitions. Being unveiled were a new generation of kinder, gentler anti-personnel devices: self-deactivating landmines, which could be set to turn themselves off after a predetermined number of years. Landmines with in-built notification systems, which could be remotely activated in the event of a peaceful, happy conclusion to a conflict. Aaaaw. One firm, with a cynical audacity I couldn't help but admire, had even developed a device incorporating seedlings, so that within a year of its interment, an identifiable blooming plant would sprout above the mine. The local populace could then be shown pictures of it, and informed not to try picking this particular flower in case it took the huff and blew their legs off in retaliation.

However, resourceful as their engineers undoubtedly were, these people were always on the look-out for new innovation, or even better: new ways to sell the same old shit. Having gathered a veritable harvest of business cards on an incognito visit to the show, I contacted several executives on their mobiles and, after establishing convincing credentials, invited them to a top-secret demonstration that would, using only existing, pre-Ottawa technology, send shockwaves through the industry. 'Once you have seen this,' I promised, accurately, 'you will never have to worry about treaties, sanctions or embargoes ever again.'

Six of them turned up, representing four different companies, each disappointed to see the others because this meant it was likely to be a rights auction rather than a free chance to get a jump on the opposition. I had made them give various assurances that they would not divulge the time and location of the demonstration to personnel from other firms, a bluff that they were understandably miffed at having swallowed and largely adhered to. Nonetheless, nobody turned back: with their competitors showing up, nobody could afford to miss out.

The location was a concrete agricultural outbuilding on some Lombardy farmland. I leased it from the farmer a week earlier, killing him the day before my show. Collateral damage, as the Yanks say.

I greeted the six delegates, fielding a few grumbles about misleading them as to the exclusivity of the demonstration. I assured them that exclusivity would not be an issue: this was something for the whole of the industry, not any individual or any single firm. 'As you are about to understand, there will be plenty to go around.'

I ushered them into the building, holding open the sturdy steel door which I had fitted to replace the aluminium lightweight effort that had been there originally. Once the last of them was inside, I closed it behind them, slid the reinforced-steel bar into place and padlocked it. As there was a table inside, generously laid out with champagne flutes and several ice buckets containing bottles of Veuve Clicquot, none of them were perturbed by this, if they even noticed.

I then made my way to the far end of the building, where I let myself in through the other door, climbing on to an observation gantry upon which stood a video camera on a tripod. The champagne-sipping delegates were gathered on a narrow strip of concrete approximately forty metres away, and between them and myself was an earth-covered area of ground, extending the full width of the building. They were all standing at least a foot back from the edge of the concrete, for fear of smearing their uniformly expensive footwear. No trip to Milan was purely about business, after all.

'Good morning, gentlemen, and thank you all for coming,' I announced, bringing them to order. 'I apologise if I misled anybody as to the uniqueness of their invitation, but I promise wholeheartedly that I was telling nothing but the truth with regard to the demonstration in which you are about to participate.'

(At this point, one or two were no doubt assuming these last few words to be either merely clumsy phrasing on my part or a mistranslation on theirs.)

'All of the hardware involved is pre-Ottawa, manufactured and sold by companies represented here today, and once you have experienced, first-hand, what it can do, I guarantee you will no longer be concerned with treaties, sanctions or embargoes. '

I produced a remote from my pocket and pushed some buttons. Several red lights blinked into life along the back and side walls. A few delegates proved themselves sufficiently familiar with their own catalogues as to recognise the devices the lights were mounted on. An incredulity borne mainly of cosseted smugness prevented panic: 'We're important and successful—nothing that bad ever happens to *us*.'

'As the two delegates from Ordnance Systems Europe will be no doubt proud to inform you, the devices I have just activated are modified versions of their own best-selling Shashka AP, altered to explode on a timer, which will run out in six minutes in the case of the mines behind you. The devices lining the side walls will explode in parallel pairs, staggered by two minutes from your end to mine. The door you entered by is secured by a deadbolt-and-padlock combination that would require a solid ninety seconds of oxy-acetylene work to cut through, if there was anyone out there inclined to assist and who was conveniently thus equipped, which there isn't.'

At this, one of them scuttled over to the door and gave it a trial push, then a more desperate full-blooded shoulder barge once it struck him that this might be for real.

'The Shashka APs, as your literature proudly boasts, have a kill radius of twenty feet, so you really don't want to be hanging around up that end for too long, no matter how good the champagne. The good news is, the door I just entered through will remain unlocked for you to exit at any time. The bad news is, between it and where you're standing lies buried a veritable compendium of your various products. It is the landmine equivalent of a jumbo-size Christmas selection box: the finest that OSE, Gieselcorp, CMK and BDE have to offer. There's Limpets, Katzbalgers, CMK-13s, Razorclams, Shurikens, you name it.'

The cosseted smugness effect was merely reinforced by my outlining their predicament. The more I talked, the more convinced they became

323

that the Shashkas were blanks or replicas. There were even a few tuts, as it was assumed they were being subjected to some kind of protest, and they were already pulling on their ideological flak jackets. Their body language mostly said 'yeah, yeah', expressing not fear, but anger that they'd been had and been grossly inconvenienced by some sandal-wearing bleeding-heart peacenik.

If how wrong they were was a number, it would have to be written down in scientific notation.

'Okay, yes, very good, we get your point,' ventured one of them, a Gieselcorp suit named Hans Mueller. 'The evil hypocrites of the arms industry who never have to face the threat of their own products, something like that?' he suggested in his heavily accented English. 'I see you have a video camera to record our discomfort and show your friends, perhaps even put on a website, yes? But here in Italy there are very strong laws regarding false imprisonment, yes? So even bringing us here under false pretences could see you face a custodial sentence. It would therefore be in your own interest to surrender the videotape. Because though it could be used to embarrass us, it would be far more damaging to you. So I suggest you stop recording and hand over the tape. You can still tell your friends your story about how you shamed and defeated the evil businessmen. No proof, but no need to go to jail either.'

The guy hadn't left the concrete strip, but I believed this was still out of concern for his shoes.

'You want the tape, come and get it,' I said.

He stared up at me for a moment, muttered something to the other suits and then shrugged.

'On second thought, keep the tape. It's

supposed to show us cowering in fear or hammering at the door in panic. Not quite so useful to your protest campaign if instead it just shows us ignoring this infantile charade and . . .'

I'm guessing he was going to say 'walking to the other side', which is what he was in the process of doing (hang ze Italian lezzer, zis is about *principle*) when he got blown so high that his body hit the corrugated ceiling before disintegrating on the way down.

Have to hand it to the manufacturers: the integrity of their trigger construction was superb. Only one of the subsequent explosions caused a nearby device to detonate without being stepped on. At first I thought it was two, but when I replayed the tape in slo-mo, I was able to determine that it was actually the weight of a CMK exec's head, descending from thirty feet, that had activated the previously suspect AP.

If you're interested in these things, out of national pride or corporate profile or even just spread-betting, it was a Dutchman from OSE who made it furthest, almost halfway, in fact, before he realised he'd stood on one. It was a limpet with a weight-release trigger, so he remained rooted to the spot until the flanking Shashkas got him. I've often wondered what was the last thing to go through his head. My best guess is his arsehole.

*   *   *

What do you mean you never heard of the landmine-merchants massacre? The atrocity that drenched the defence industry with a bloodstain it was forced to launder in public? That stimulus of international debate and bitter controversy, every condemnation of the murders nonetheless dripping with inescapable equivocation due to the victims being hoist by their own petard? How could you possibly have missed a story as big as that?

Exactly. They moved heaven and earth—mostly earth—to cover it up, with God knows what levels of collusion and influence being brought to bear. It didn't happen. Not so much as an internet rumour. Even the murder of the farmer got the screens thrown up around it: officially attributed to some travelling indigent.

By way of consolation, the experience served me well for the future, teaching me that the events had to be immediately accessible to the public, after which there could be no way of the authorities getting the toothpaste back into the tube. At the time, though, it was more than a disappointment at my wasted effort: it was a humiliation, because of what it taught me about my own predicament.

I had the videotape, of course, but there was the rub: I was scared to release it. I reckoned I could dub over my voice and edit out any glimpses of myself from the footage, but I didn't know enough about the forensics of these things to be confident I wasn't giving anything away. My humiliation lay in that I had too much to lose by drawing so much attention to myself, and it wasn't just the authorities that I had to worry about bringing down upon me. No matter how smart or careful I thought I could be, the second I announced myself, I was starting a countdown to my own death.

That inescapable risk-benefit equation: if I didn't identify myself as the author of the deed, it was useless by way of altering my son's legacy; but to identify myself as the author of the deed was suicide.

I had to make do with trips to Scotland to see my son as close up as I dared. The double disguises of being presumed dead and of no longer quite resembling the deceased anyway served me well. I did, however, unavoidably spend a great deal of time in hotel rooms, with little to pass the time but British television. I got very, very close to my boy on occasion, though never enough to exchange words, to look into his eyes while we spoke to each other. Over time, the frequency of my trips increased, as did my frustration at what I couldn't do for him, and my awareness of what I must. The idea didn't go away: the sense of untapped power and possibilities grew and grew. Only the price never changed.

*   *   *

Physicists ought to look into the mass-gravity relationship as it is uniquely warped by dead weight. No matter how skinny and waif-like the body, it always seems to feel 50 per cent heavier once it's lifeless. This girl is five-foot nothing and, I'd estimate, seven stone soaking wet, yet she still has me worked up into a sweat as I get her ready for transport. Could be my condition, right enough—mustn't forget that.

Even just sitting her up while I change her t-shirt takes some haulage. I am grateful, as ever, that once I get the body inside the golfing flight-bag, it will be a lot more easily manoeuvrable. Who invented these things? I ought to look him up and

327

send him some champers or whisky by way of gratitude. I don't know how they rate for taking your clubs abroad, but for moving bodies around—in broad daylight, if need be—they are second to none. Even the golf bag itself, once you remove the internal dividers, makes an ideal means of keeping the stiff—or more pertinently, I should say, the flopper—in place and the shape disguised. You zip the canvas flight-bag around the whole affair and then you can just wheel it about on the rollers.

Her old t-shirt is a bit whiffy, but that is understandable after what she's been through. We'll never be in a hurry to add scent to sound and vision on our tellies or the internet.

I begin by trying to haul it over her head, then give up and cut it with a knife. Pitiful to think there had been paparazzi stalking her in vain for weeks, tabloid picture editors on tenterhooks for the thus far unattainable prize of a topless shot, and internet Photoshop geeks doing digitally manipulated zoom-ins of wait-was-that-a-nipple? frame captures from her *Bedroom Popstars* dance routines . . . and all because of these two skinny tits.

I pull the new t-shirt over her head and tug her arms through the sleeves, then stand back and take a picture, in case it gets suppressed or its significance missed. She is slumped in the shot, her head lolled to one side and her hair hanging partly over her face, straggling ends reaching on to the slogan. Has a certain zombie-chic about it.

I wheel her out to the van and check my watch. Couldn't be late: have to catch the last train, after all.

I make it to Graining North station for twenty

328

past eleven. It is an unmanned facility, on the outer limits of the tube system. Network rail trains thunder obliviously through it, with even the local mainline services stopping at Graining West instead. I sit inside my vehicle in the car park for a while with my lights off, keeping watch for any other approaching passengers. It wouldn't matter if one showed up once I was on the platform—I'd just make sure I was at the very end—but I don't want anybody coming through the station as I make my way from the van.

I pull my hood all the way forward and keep my head down from here on in. I'm not quite ready for my close-up, and I'm certainly not giving them it on the shitty CCTV system that's monitoring the unmanned station.

There is nobody in sight, and nobody already on the platform. Would have been unlikely anyway, as I timed it to arrive in the car park shortly after the penultimate service departed.

I stand and wait for an irritatingly long six minutes. The final train is four minutes late, just long enough for me to worry that it has been cancelled, that they've changed the timetable or the lazy cunts have just decided to call it quits early. However, at 11:38, along it comes, the final service into central London, originating from the Graining West terminus one stop up the line.

I watch it rumble arthritically along the edge of the platform. As it passes I see only one passenger, seated with her back to me. I roll my companion along behind me for a couple of dozen yards so that I can board several carriages along.

Unsurprisingly, the other passenger doesn't alight. Who takes the tube a solitary stop from

Graining West to Graining North, especially at this time of night? Nah, she is heading into London: probably on her way to a night shift somewhere, emptying bins and wiping down desks before the daytime worker drones return to don their headsets, boot up their PCs and jack their souls directly into the system that controls them. No wonder Sally here had hurled herself at every camera rather than take her place among the indentured undead.

Oh yeah, forgot the drum roll and trumpets. The winner of *Dying to be Famous* is . . . Sally. Yay. There'll be more of a fanfare when the public finds out. To which end, I get to work once the train has pulled away.

I unzip the canvas and haul her out of the golf bag within, lying her carefully down on the frayed upholstery of the seating. Then I zip the flight-bag again and make my way into the next carriage, in case anybody gets on the one she's in at the next stop.

Nobody does. It's another empty station. Nobody but me gets on the service heading back to Graining North from the opposite platform, and nobody else gets off there either.

As I walk briskly back to the van, I wonder which unsuspecting traveller will win the golden ticket to their own fifteen minutes by being the one lucky enough to find her. It's a limited offer, though: as she's in the end carriage, she could make it all the way into central London before she's disturbed, so there's a good chance she'll come to on her own and stagger off the train herself.

For she's alive—*alive!*—say it in your best Bride-

330

of-Frankenstein voice. Alive: that's the prize; part of it, anyway. Accompanied or not, very soon she'll walk off that carriage, blinking and disorientated, into a sea of gaping faces and a hundred clicking camera-phones, all trying to seize their own slice of the moment. And this will only be the beginning.

Congratulations, Sally. You don't belong to me any more—but that doesn't mean you're free. You belong to them now. That's what you never understood. None of you did.

And won't everybody in the whole country be so happy to see her? A good news day at last on a story that had only produced horror and heartache. A glimmer of hope for the cops too, a chance, finally, to get some first-hand information. Yes, they're all going to be happy, so very, very happy that Sally has been found alive.

It will only be later that they realise this was the opposite of what they ought to have wished for.

PUSH THE GHOSTS

'More head-hunters in looking for you,' Morrit says. 'Asians again, I think, or maybe Middle East: Dubai and that.'

The old man manages to make it sound like he's a lonely room-mate passing on phone messages from numerous girlfriends. There's an aridity and affected grumpiness about him that can make anything he says sound like a complaint. Took Zal a while to understand that it presented no reliable indicator upon Morrit's state of mind: it was, as Lizzie had helpfully explained, merely symptomatic of his being from Yorkshire. He takes a singular satisfaction in sounding pissed off even when inside he is whistling. Today, though, his weary tone reflects a frustrated sense of futility about the information he is conveying. No matter how enticing the potential engagement sounds, there is no point in Morrit pondering the prospect, he believes, because Zal will inevitably reject it.

It would be inaccurate to say that it has become a growing source of tension between them, but it has definitely been leading to an increasing sense of puzzlement and confusion on the old man's part, and Zal is aware it will soon have to be addressed. They are coming to the end of another contract, and they both know it's time for a change. The confusion part stems from Zal having concealed that he concurs, thus Morrit is wondering if Zal is planning to just stay on this ship forever.

Their first contract was for six months. Six

months had turned into four years and Morrit couldn't understand why Zal didn't appear to have itchy feet. Given that Morrit referred to him as 'Mac' and believed his name to be Innes McMillan, it would be fair to say that Zal had never entirely levelled with him.

Lizzie had left about a year and a half back, to get married. There weren't a lot of single males on these trips—not ones under sixty, at least—but she had met this guy who was a senior executive for the cruise line. Morrit and Zal both knew she hadn't intended to be extending her career as a magician's assistant: she had stuck with her dad when he needed her, knowing he would be forced to retire from the game soon enough. But then Zal had come along and changed everything, and she found herself still climbing in and out of boxes as her thirties trickled away. They were both sorry to lose her, but Morrit took a deep satisfaction not just from her happiness, but from no longer worrying that she had sacrificed too much of herself just to look after him. Unfortunately, this left him free to channel all his paternal concerns, as well as his guilt-stained gratitude, towards the overall welfare of Zal instead.

Morrit's arthritis had slowly but inexorably worsened, to the extent that he eventually had to accept that he would no longer be able to usefully grip a saw, never mind anything more precise. Over the same time, however, he had been teaching Zal the very skills he was losing, and so was able to direct his apprentice's hands in realising the props and apparatus that they had collaborated to design. He was a living encyclopaedia of both hardware and technique, which, allied to Zal's stagecraft, made for a very successful collaboration. In the absence of Lizzie, they had put their efforts into designing new tricks for Zal to perform alone, rather than into looking for a replacement. They liked the way they could spark ideas off each other and liked the challenge of reshaping the show, but another unspoken factor was that they weren't ready to let an outsider anywhere close to their professional partnership.

Morrit was so grateful for this Indian summer to his career in magic that he tended to forget how invaluable he was to Zal, and that was what fuelled the recurring thought in his white-maned head that the younger man ought to be spreading his wings.

'You shouldn't be stuck on this old tub at your time in life, lad,' he had said a few months back. 'You're too good for this.'

'I like it here,' Zal replied, neglecting to elaborate by way of indicating that he didn't wish to discuss why.

'I like it too, but I'm an old man. You've got too much more you could achieve. Your name should be in lights somewhere. You could pack 'em in wherever you went: London, New York, Vegas, anywhere.'

'I'm content to just be for a while, Dan,' he told him.

Morrit nodded and frowned, understanding that Zal was alluding to things he wouldn't talk about: 'Never ask me and I'll never need to lie to you,' had been how Zal once closed the subject, way back on the night they conspired to dispose of Fleet.

'Aye,' Morrit said, sighing, 'but remember a ship

334

is just a means of transport. It's only supposed to take you somewhere else. In mythology, a ship usually symbolises a journey between the worlds of the living and the dead. I don't know what it was happened to you once upon a time, lad—I've learned not to ask—but I know you're not in a hurry to rejoin the land of the living, and I'm just saying: you want to watch you don't end up lost in transit.'

Zal couldn't entirely put this down to Morrit merely transferring the way he used to worry that he was holding back his daughter, as the old man had stated his readiness to follow Zal wherever opportunity might take them. Despite pushing seventy, he had no reluctance about the prospect of, for instance, relocating their show to some air-conditioned oasis in Dubai.

'There's no retirement for the likes of me. Wife long since passed, Lizzie moved on. Got a few bob put away now, thankfully, but when you retire, it's supposed to be to do what you like. *This* is what I like, what I love. It's all I know. I'll do it until I die, and if I die doing it, I'll die happy.'

Representatives were booking on to single legs of the *Spirit of Athene*'s route just to watch 'Maximilian' perform, and sound him out about future engagements: cruise ships from the Caribbean to the South Pacific, hotels from Dubai to Tokyo. Four years of non-stop work under the tutelage of Daniel Morrit had turned Zal into an excellent magician, and given the money they were offering, those Asian and Arabic reps clearly shared the old man's belief that he could pack them in wherever he chose to go. Zal reckoned the 'name in lights' part of Morrit's prediction was on the money too, and that had been the problem. He had grown used to the security of alias and anonymity. He had created a small world for himself that he could guard and control. The mere prospect of publicity reawakened old fears. It had been a long time since he saw off Albert Fleet, but if his face ended up in a magazine, a brochure, some TV promo spot, then it increased the risk that his past would once again catch up with him.

Tonight, though, Morrit's weariness of tone is misplaced. Zal doesn't tell him, but he is ready to audition for the Arabs, the Asians or whoever has been asking for him. He's been ready for a while. He'd enjoyed a rehabilitative period in the comfort zone, and had spent his time both profitably and wisely, but he'd always known he couldn't remain in permanent stasis. The old man was right: it was time for something more.

He had first set sail on the cruise ship in order to spare Angelique, but found himself developing an invaluable relationship with Morrit: that was who he now feared they'd go through, like they went through Parnell, his prison-time mentor, like they went through Karl, and like they went through his dad. As long as you cared for somebody, you were vulnerable: if they always hurt who you loved, then either you couldn't afford to love, or you couldn't afford to let the world hear your name.

But who the hell wants to live like that?

He thinks of Fleet, thinks of the Estobals, thinks of Bud Hannigan. They had all taken him on and lost, then taken him on again and lost more heavily second time around. The bounty hunter still knew where to find him, and yet nobody had come calling in four years. Perhaps the penny had finally fucking dropped. Perhaps the ghosts of his past had more to fear from him than the other way around.

The houselights go down and he takes the stage. He's buzzing, sensing that tingle in his fingers as he touches the cards. He feels supreme, determined to give his best performance but not remotely nervous about the stakes. Four years of this, of learning, practising, improving. He is at absolutely the top of his game, exhilaratingly so, and nothing, nothing whatsoever, could faze Zal Innez right now.

Apart from seeing Angelique de Xavia sitting in the front row.

The magician's hands suddenly spasm as he grips the pack. The cards explode from the collapsing cradle of his fingers, spraying, spinning, fluttering about the stage like crisp autumn leaves stirred from the gutter by a sudden gust. It is not a flourish but a fumble, a moment of startlement. A trick derailed, an unscripted incompetence. Some members of the audience gasp, others fail to stifle giggles. The muted laughter is horrible: a cringing combination of being embarrassed on the faltering magician's behalf and being embarrassed by being present at such a tawdry spectacle. But can he recover, that's the question? Does he have an out?

Does he hell.

SHOULD MAYBE HAVE WORN
HER JEANS

'You were godawful,' Morrit tells him when he comes offstage. Zal reckons he is being a little harsh, but decides to consider it a compliment regarding the standards he's set himself.

He had been forced to laugh off the involuntary card scramble precipitated by seeing who this supposed Asian talent agent really was, referring to the calamity as 'an old trick known as the fifty-two-card pick-up'. After that, he didn't actually blow anything, but though it would have taken an eye such as Morrit's to detect precisely how below-par he was, his performance was undoubtedly lacking in spark. It wasn't that he was simply going through the motions of his act because his mind was on other things; rather, it felt like going through the motions of his act was the only thing he knew *how* to do right then, because inside his head was one huge fifty-two-card pick-up. There *was* a moment, only a glimpse, during which he almost connected to the buzz he should have been getting from doing this act in front of Angelique, but it was immediately swamped amid questions, fears, doubts and confusion, just one card in the wildly buffeting pack.

'Not exactly what I'd call an audition performance,' Morrit goes on. 'I'd say I hoped it didn't bugger up your chances of any offers if it wasn't that it only took me about ten seconds to deduce that the young lady seated at table four most probably wasn't here on the business of talent

338

recruitment after all.'

'Fair to say her specialty is less recruitment than . . .' Zal cuts himself off. 'No. She ain't here about a job. Not a new one, anyway.'

'That wasn't all I was able to deduce. Your picture is finally starting to take a more recognisable shape, with that particular piece slotted into the jigsaw.'

'There's still a lot of holes, Dan, and believe me, the end result's an optical illusion.'

'You think there's an illusion ever been constructed that I can't see through? Don't kid yourself, lad. She's why you're here, and why you stayed here. She's why you never got involved with anybody all these years, more than the odd casual fling.'

'We don't get a lot of hen parties sailing with us,' Zal suggests, but that was never going to head Morrit off. Zal had been made, big-time, in front of a full-house.

'I've watched you closer than you think, especially since Lizzie moved on, as I've had nowt else to fuss about,' Morrit tells him, helping Zal off with his jacket and delicately removing the aluminium card-pull apparatus from his back. 'There's a big crew on this tub, lot of shipboard romances, but not for you. A one-night stand here and there maybe, but nobody gets allowed any closer than that. This boat's your Foreign Legion, lad, the good ship *Beau Geste*. Who is she?'

Zal sighs, turning to look his partner in the eye.

'You remember when I said never ask me and I won't need to lie to you? Well, she's everything I'd have to lie about.'

Morrit holds his gaze for a moment, weighing

this up with piercing scrutiny. His face softens and he nods. 'Just like always, lad: you've told me everything and nothing.'

'Best I can do right now,' Zal says sincerely, to which Morrit nods again. 'Where is she? Did she leave for the bar, or . . . ?'

'Still in her seat, last I looked. They'll be ushering her out in a minute, if they haven't already.'

'Love to see them try,' Zal mutters. 'Can you go get her? Bring her here?'

'*Inside* the sanctuary, sir?' Morrit jokes, putting on a Hammer Horror voice.

'Access all areas,' he confirms.

'Very good, sir.'

'She's had that since day one,' Zal says quietly to himself, after Morrit has exited.

It takes Morrit an age to retrieve her: maybe as much as ninety seconds. Time enough to revisit several years' worth of conflicted thoughts, and to reflect on the reality that while he may once have made a decision, there has never been a resolution. He needs more than ninety seconds to prepare for this, but if five years hasn't been enough, then maybe it is as well there isn't any longer to worry about it.

He couldn't see her closely from onstage: merely close enough to have no doubt that it was she. The old man leads her inside, holding open the dressing-room door, saying nothing. Christ, what could the guy say by way of introduction, on behalf of either of them? Morrit then withdraws from the dressing room with the delicacy of some eighteenth-century servant, stepping backwards with his head low, but breaking below-stairs

340

protocol at the last to snatch a parting look at Zal. What perhaps started as curiosity is now bordering on concern. Zal returns a neutral look as Morrit closes the door. Then, finally, after five years, Zal Innez and Angelique de Xavia are alone together once again.

Zal is leaning against the table-top that sits hard against the wall below a bulb-rimmed mirror. Angelique stands just inside the door, clasping her hands in front of her dress. There is something supplicant about it: she looks like she wants to fold her arms but is resisting, trying to appear more open.

They just stare at each other for a moment or two, both mouths open a little but no words finding their way forth. He takes her in, close up and in full light. She looks thinner, shorter, delicate. Smaller than he remembers, but that's because his mind can't accept how so much—all the things she came to mean to him, all the memories of who she was, all the women he imagined her now to be— could possibly fit inside this one female frame. She seems sharper-featured, though this also could merely be the effect of the reality suddenly superseding the memory and the ideal; it could also reflect that few of her days since he last saw her have been spent tending baby lambs in an Alpine glade. She looks tired and uncharacteristically vulnerable. They're standing here in the civility of a cruise-ship performers' dressing room, and yet she looks less sure of herself than when she was being held at gunpoint. (By him, he feels a little ashamed to recall.)

Zal speaks first, forces himself to come up with something to say. He's the host here, for one thing,

341

but for another, he has this crazy feeling like if he doesn't reach out and make contact somehow, then whatever spell summoned her here will be broken and she'll vanish before his eyes.

'I have to compliment you on your timing. I think I finally got over you about eighteen minutes before I went onstage.'

That was the first thing that came into his head; just a line, but borne of too much truth and sincerity for its thus jarring flippancy to get past his tact filters. He doesn't say it. The look in her eyes didn't suggest she was in the mood for sparring: her deflector shields were definitely down.

What issues from his mouth instead, at shorter notice, but with even more truth, and even more sincerity, is simply this:

'I missed you.'

She still says nothing, her mouth remaining slightly open and a look of uncertainty persisting in her eyes. The moment endures without her forming any kind of response. Zal wonders, for just a flash, if she's looking so conflicted because she's here on Judas duty and she's just about to arrest him. His words on her doorstep that last night, when he effectively put himself at her mercy, repeat in his mind: *You can make this my Gethsemane if you want. It would be worth it for the kiss.* He realises right then that he still feels the same as he did then: that he wants that kiss, wants to be with her more than he fears for what he has to lose. He only ever rationalised otherwise because she wasn't there.

He hears a sound of her breath, a cut-off sigh which he realises was her abortive attempt to speak the word 'I . . .' She closes her mouth, swallows. He thinks she's about to give it another shot, but she doesn't. Instead she takes a pace forward, her eyes filling, and throws her arms around him.

Should maybe have worn jeans, she muses to herself with a slightly embarrassed giggle. That it's the first giggle she has emitted in weeks tells her that, actually, maybe the dress was right.

She joins the ship at Barcelona, where it is docking overnight. As soon as she boards, she goes directly to the ballroom where 'Maximilian' is billed to appear, not even stopping at her cabin on the way. She is told by a mercurial-looking older man that 'Mr McMillan' is in preparation and cannot be disturbed until after the show. She suppresses a smile at the mention of this new alias, Zal's middle name, given him by his Rangers-daft exiled father in tribute to his favourite player. The older man's name is Morrit, and she deduces immediately that he is no mere shipboard functionary, even though he is using precisely that posture to make out he is powerless to assist. Even over a brief exchange, she can tell he and Zal are in league somehow. Like with Zal, she gets the impression there are several discrete but interconnected levels of activity going on behind this guy's eyes: secrets, schemes and mysteries that would continue getting on with themselves quite independently while his external manners dealt with the mundanities of the here and now.

It would thus not, she reckons, be worth getting pushy.

Instead, she accepts it as for the best. She will go back to her cabin, unpack her bag, freshen up and go along to the show. That way she can take him in gradually, from a safe distance, like an immunising dose, before confronting him point blank.

She's travelled light, only thrown a few things together in hand luggage. Barcelona is warm, below decks on the ship warmer still. She opts for the dress because it is light and lets the air get around her, whereas her jeans were starting to feel that heavy and clingy way. She only realises once she has showered and slipped the dress over her head that she hadn't even been thinking about which outfit she looks better in. It is a strangely jarring occurrence: the single-minded and pragmatic motives that are both driving her and guiding her have suppressed such considerations as unhelpful and distracting, but just this one thought seems to shatter the seal. Thus far it has all been arrangements and logistics: the ship's schedule, onboard entertainment programmes, flight timetables, the pursuit and practicalities of putting herself and Zal together in the same room. How she is going to feel about that and how she is going to conduct herself when it finally happens have been subconsciously deferred as not of immediate concern, like they were arriving on a later flight. That plane has landed now, and she is glad Morrit sent her away and gave her time to prepare.

She looks at herself in the mirror and begins wondering, for the first time, what Zal will see. She appears tired, worn-down, fraught, like she's barely slept in weeks, which is about right. Does the dress make her look like a scrawny waif? She has a spare top, one with long sleeves. Maybe in that and the jeans she would look . . . No. The pragmatic drive kicks back in. She can't afford this. Go with what's physically comfortable. The show's starting in half an hour, and you want a table near the front.

When the houselights dim, Angelique is more anxiously expectant than she remembers feeling at any teenage rock concert. She feels a surge throughout her whole torso, something horribly involuntary, a reminder of how much we are hostage to the physiology of our animal selves. Jesus, she's never felt this helpless entering the line of fire on Dougnac's team, but that's because under those circumstances, she's always known what she was doing. Then all of it is suddenly dispelled and forgotten as Zal Innez takes the stage, dressed in a sober but semi-formal dark-coloured suit and bow-tie.

He looks more mature, rather than simply older. The lighting and a bit of make-up are no doubt contributing to this effect, but it's the fact that he's lost the long locks and the peroxide that is giving him an upscale elegance that is almost impossible to equate with the prison-tattooed Californian surf punk who stole from her bed as she slept on Christmas Eve 2002. Some things can't change, however: the blue eyes are still two piercing laser-beams, and even from the rear of the stage, she's well within their range.

She had imagined he'd go for a magician-as-rock-star look, though hopefully not including a David Copperfield retro-nineteen-eighties mullet. Instead, he's closer to retro-eighteen-eighties, and looks quite disarmingly respectable for it. The whole set, in fact, looks defiantly old-school and unashamedly low-tech. She interprets it as a statement of embracing the past, though that might just be an optimistic personal spin.

Angelique feels a reprise of that involuntary surge as he comes to the front and walks across the stage, ever closer to her, a deck of cards in his hand. Will he look her way, she wonders, and if so, will he see her? The lights are bright, and even though she's at the front, she may appear no more distinct than a silhouette. Besides, with his evidently practised stagecraft and effortlessly polished professionalism, he's unlikely to skip a beat, so how would she tell either way?

He fans the cards, turning a little on his heel, and scans the front row.

Yeah, she's pretty sure he sees her.

He recovers from the fright, albeit not exactly seamlessly, and proceeds to keep Angelique spellbound for much of the next hour. She forgets where she is several times, lost in just watching him perform; maybe even in just watching him. There's a wrenching inside each time she remembers the reality, but soon enough she's lost again. He looks happy. He looks like he belongs. He is doing what he was meant to be doing, she can see that, and wishes she felt the same way; has been wishing it for several years now. She thinks of talking to him by the Seine, what he told her about his father, how he rejected what his dad wanted for him: rejected all of this. She looks at the set, remembers her first impression: embracing the past. Embracing *his* past. Reconciled to himself, reconciled to who he ought to be. What can she possibly do here but jeopardise all of that?

Then, once he has taken his bows and she is left there, alone at her table, she finds herself thinking only of why she has to.

When she pictures her mum and dad, she doesn't envisage them as they were the last time she saw them, or before she left for Dougnac's job in Paris. She sees them as she did growing up, when they seemed larger, more vivid, the most dominant force in her life. They drove her crazy, but only because they loved her so much. She used to think James was the apple of their eyes, that he could do no wrong, was extended far more freedoms, granted so much more slack. She, by comparison, never seemed able to do enough right, and she felt like they were always on her case, keeping her under permanent surveillance. Hence her having to lie about going to the cinema so that she could sneak off to watch Rangers at Love Street, while James had free rein to travel to Aberdeen, Dundee or Edinburgh to watch the Tims. Hence James's choice of profession was a source of pride and hers was a source of disappointment if not outright disapproval. Being a lawyer was something they could relate to: her dad had been one in Kampala, after all, before having to settle for the reduced—but welcome, to a refugee—circumstances of a clerk in a tax office. To two people expelled by Amin with one child in arms and another on the way, however, the position of police officer carried no such upstanding connotations. To them, cops were paid thugs, bullies, sneaks and cheats. They enforced laws but did not make or even debate them, thus they were, even given the benefit of the doubt, a lesser profession, for lesser people. Then, of course, there was the small matter of that paucity of men 'on the horizon', and the even more conspicuous absence of tiny feet pattering somewhere in their wake.

She thinks of these things now because they, more than the easier times and the happier memories, bring home how much she was loved. They fussed, they protected, they fretted—they bloody well *surveilled*—all because she was so precious to them. They wanted it all to be perfect for their little girl, their wee princess.

She remembers being held in her dad's arms, how no matter her age, it would still make her feel like she was three years old and that the world was full of love, care and security. These past weeks, she has felt as bereft of love, care and security as she has ever done in her life.

The slight and soft-footed figure of Morrit approaches from the side of the stage, smiling coolly, politely, nothing overbearing. He says 'I'll take you to see him now,' and offers her a hand up from her seat. It's like he already knows this is going to be delicate. He leads her backstage and down a short passageway. She tries to think of what she ought to say but her mind is just white noise: never mind words, there are no images, no memories, no thoughts whatsoever.

She is barely aware of walking through the door; it almost feels like she has disappeared from her seat and rematerialised inside the dressing room. She is definitely unaware of Morrit leaving; the guy simply vanishes, leaving her standing a few short feet away from Zal, close enough to smell his cologne mingled with the fine, fresh sweat of his exertions under the lights.

They look at each other a moment. His bow-tie is undone, his top few buttons loosened, the tops of tattoos visible on his chest: just slightly dressed down, it's enough to make him look entirely the man she remembers.

She knows she won't be able to speak. She doesn't know what to say, how to explain anything: from why she wasn't at the Musée d'Orsay to why she *is* here now. She glances down at herself, her skinny, suffering form in this travel-creased old dress, and feels suddenly wretched, wonders how she could have deluded herself that whatever happened between them was enough to justify her being here, that any spark of it could remain in this man all these years later, far less enough to ask of him what she must.

And then he says: 'I missed you,' and it sounds like everything she needs to give voice to, condensed into three simple words.

She tries to reciprocate, but even 'I missed you too' is three words too many; she's not even sure if she managed an audible 'I'. She wonders can she try again, but already feels herself fill up. There can be no words now. She steps forward and throws herself against him, just as she's throwing herself on his mercy. His arms enclose her, the most welcome weight upon her shoulders. She buries her face in the open part of his shirt, as though it will stifle her tears, prevent weeping becoming full sobs. It does, just about. She feels his head bow down, his nose against her head, breathing in the smell of her hair. Then one of his hands cups itself around her cheek, tentatively testing whether she's ready to lift her face again. She isn't, quite, but she wants to kiss him more than she wants to further stem her tears.

She lifts her head and kisses him tenderly,

delicately, feeling the sobs subside and even the tears dry as she closes her eyes and loses herself in his taste, his scent, his touch, the sensation of being pressed against him. Then in the instant their tongues brush together, something is ignited, and delicacy is no longer enough. She kisses him deeply, insistent beyond desperate, and feels a popping sensation close to her breast as her hand tugs his shirt open a little too impatiently and tears off a button. She places a hand on his chest, but feels her wrist still snagged by material, so she puts her hand inside his waistband, intending to haul out his shirt. As she does so, her fingers brush against his cock, at fatefully the same moment his hand rides up the back of her leg and cups her right cheek through the M&S cotton. There's no going back from there. Not in this dress.

She pulls him down to the floor, lays him back and takes him inside her. After only a few seconds, she feels him spasm. He's lasted a lot longer than on their first encounter, but it's not enough. Not this time, not now. Maybe if she'd given into the sobs, but she didn't. She takes his hand, guides it between her legs. He's ejaculated but he's still inside, still hard. She needs this, needs the release, needs any kind of release, and his hand, his wonderful hand, his taste, his smell, the feel of him against her, the feel of him inside her, gives her that. He makes her come. As the song—their song—prophesied, he will *always* make her come.

She lies there on top of him for a while, breathless, reeling, his hand stroking her hair, her cheek against his chest. Then, when she feels ready, she turns her head and looks him in the face.

Her first words to him in five years are a sheepish: 'Is there a bathroom round here?'

Zal nods to a door at one end of the room.

She climbs off him and scuttles away, grabbing her so easily discarded knickers from the floor as she passes them. Should maybe have worn jeans . . .

*   *   *

'So, how you been?' he asks, the casualness of it his way of acknowledging the chasm since last they spoke. He's inviting her laughter but its onset feels like it will trigger more tears, so she reins in both.

'I've been better,' she states. 'I'm in trouble, Zal. I'm in a lot of trouble.'

He nods, like it's no great surprise. 'Well, I brought you plenty back in the day. Only seems fair you should take some to me.'

He's smiling, looking sympathetic, inviting her to talk, but as ever, it's hard to know what else is going on in there.

'I hate to bring this to you. It's not fair to bring it to anybody, but you're the only person I believe can help me. I wouldn't have come otherwise.'

He smiles again, but there's a sadness in the corners of his mouth, something even he isn't quite able to hide. It speaks of regret, disappointment. Angelique knows he'll help, but he knows it may cost him.

He squats down at a knee-high fridge tucked under a table and pulls it open, his fingers oscillating back and forth between bottles of beer and mineral water. Angelique indicates the purer stuff. He ignores her and hands her a chilled Dos

Equis instead, a gesture to which she can't help but smile just a little.

'Take your time,' he says. 'Talk to me. Tell me where it hurts.'

*   *   *

'So,' Zal says, when it's clear Angelique is all talked out, her bottle empty and her eyes red. 'Let me just see if I got this straight. Threatening to kill loved ones in order to coerce a resourceful individual into carrying out nefarious deeds. How on earth could that scenario have made you think of me?'

She manages a pretend shrug, but Zal can tell it's an effort. Everything is an effort. She's drained in ways he can all too readily imagine. He gave her a beer but wishes he had some elixir he could offer to truly restore her to the woman she ought to be. There's a familiar anger starting to burn inside, one long since quenched these past few years. Oh yeah, she came to the right guy, for any number of reasons. He just wishes there were circumstances under which she would have come anyway. She came because she needed him, not because she wanted him. He understands that, and knows he mustn't lose sight of it. It's okay to love her; that's not where the danger lies.

'It's not just your empathy I'm relying on,' she says. 'More your professional expertise. To me, this looks like a near-impossible situation. If memory serves, that's your specialty, but it's a big ask, Zal, a very big ask, and not one I'm sure I've any right to make of you.'

Zal thinks of the decision he made in that

apartment in Palma, when Fleet first came for him the night before his intended trip to Paris: the easiest to make and the hardest to bear. He did that for her. He gave her up—for her. He is thus able to tell her, with confidence and sincerity, if not a great deal of regret:

'There's nothing you couldn't ask of me, Angelique.'

Including taking the same decision again.

She shakes her head.

'Remember when you asked me out?' she says. 'When you called up the officer investigating the robbery you just pulled off and *asked her to meet you for a drink*?'

'The story rings a bell.'

'When I agreed, I asked you, "What have we got to lose?"'

Zal grins at the memory.

'And I replied: "Off the top of my head, everything."'

'Well, that's just it, Zal, it's different now. I'm here with my hand out because I've already lost everything and I'm hoping for a miracle to get it back. I've nothing left to lose. If you help me, you've got *everything* to lose, and don't try telling me otherwise: I saw you onstage tonight. You're happy, you're doing what you were always meant to. You've left behind all the kind of shit I'd be dragging you back into.'

'I won't lie to you. I am doing pretty good these days. I might even have said I was happy, till you walked in and reminded me what happy really means.'

'Zal, don't,' she says. She's trying to sound like she's just being modest, but he hears what she's

353

really saying: don't pin your hopes of happiness on me, because I can't help you. However, she hasn't understood what *he's* really saying.

'No, listen to me. I've lost stuff before, lost what I built up, lost out on my hopes and plans. So what? You can always build those things back up. You can make new plans, build a new life. The only thing we can truly lose forever is people: I know that better than most. Everything else is replaceable.'

And the thing I could least stand to lose forever, the thing I could never replace, is sitting right here, he doesn't add, though inside he's yelling it. It's for both their sakes. He needs her to know she won't owe him anything, can't afford for there to be any kind of contract implied by this.

'That's why I'm saying this, Zal. You could lose everything because you could lose your life. These people are more dangerous than anything you've dealt with before.'

'I've dealt with a lot of bad people, Angelique. I've met gangsters, killers, thieves, assassins, crooked cops, crooked guards. I've met men I couldn't last two minutes with if we were locked in a room together. I've met men who would kill grandmas and babes-in-arms if it guaranteed their next fix. I've met men who've ordered murders just to prove a point, and I've met the men who execute those commands. But I've yet to meet one I couldn't fool.'

At this, her eyes finally light a couple of watts. She knows of which he speaks. Their fingers interlock and she manages a sad smile: still lost, but reassured just a little.

He appreciates the danger is no joke,

understands what she's asking and knows he'll give it. He'll supply the illusions she needs. He's doing that already, and not just in his bravado: he's supplied an illusion by concealing his own feelings, his own wants and his own intentions. He'll do anything for her, and he'll make the same sacrifice he did before. He'll save her from the bad guys, and then he'll save her from himself.

THIEVES LIKE US

'So this is what it feels like to be a crook,' she says, as they sit in a 'stolen' van, parked in a secluded side-street from which they are observing the Provence Troisième Arrondissement evidence repository on the outskirts of Toulon.

'Now don't you start with that shit,' Zal replies. 'I was an *ex*-criminal until you showed up, *officer*. What kind of a cop do you call yourself? I hadn't committed a crime in five years, then I'm in your company forty-eight hours and you've got me carrying out a heist on a high-security police facility, on top of grand theft auto for boosting our ride.'

'Don't be melodramatic: we haven't stolen anything.'

'Are we, or are we not, sitting in a stolen van?'

'Taking refuge in pedantry, very masterful. I'm just saying: this is giving me a worthwhile insight into the criminal mind.'

'And you'll learn what it feels like to be in lockdown pretty soon. Just be thankful it's solitary.'

They hired the van in Barcelona and drove up overnight. Zal had reported the van stolen first thing that morning in Toulon, telling the police he had gone to grab some breakfast and come back to find it missing. The vehicle itself was not a big worry, it being a hire van and covered by insurance, but the contents were irreplaceable. While he was doing this, Angelique had been parking it in a secluded spot not far from where they sat right now.

It had been a long drive north, ought to have been a time to catch up, but she didn't feel like talking. Zal seemed to understand, though he talked enough for both of them, and she was happy just to listen.

Zal had dropped everything for her: let the ship sail without him and cancelled all shows for that leg of the trip.

'It's not like they can fire me—I'm out of contract in a couple of months anyway. Plus I've never dicked them around before: barely took a day off in four years. Why the hell would I? To travel?'

'Just don't burn any bridges,' she warned him. It was one thing to help out with the job in Toulon, but there was no timescale for dealing with Darcourt. He could (and this was the thought that threatened to paralyse her by day and steal her rest at night) simply disappear again, decide he'd had his fun. Maybe there *was* no game-plan, no grander scheme. Look how many years had passed between his previous surfacings: the Baker kidnapping, the Lombardy massacre and now his stark-reality-TV abominations. If he went to ground again, then all was lost.

Zal told her he'd decided it was time to move on anyway, though as he didn't elaborate, it sounded like another line to make her feel better. Making her feel better was his second great talent, one he was exercising to regular and very welcome effect; she just couldn't work out what was in it for him. With Zal, there was always an angle, and his inscrutability meant that you would never see it until it was too late to make your own move. Back in Glasgow, he had played her, more than once. Whatever else passed between them, she had been left with her doubts as to whether it was all part of his scheme. She knew that, in the end, his motives had been decent. She knew he was a decent man. Whether he was, more simply, a *good* man had remained a hazier question. Here, however, there was no way of playing her, no unseen motive.

The very end of that song—their song—plays in her head: a girl asking, 'Can this be for real?'; a male voice yelling: 'Noooo!'

Discreet inquiries through her French police contacts revealed that Bouviere's files were being protected by more than mere medical ethics. The late surgeon having rendered his services throughout a certain unsavoury but influential and highly motivated constituency, the clean-up of his premises and the ordering of his effects had not been left to a few lowly gendarmes and janitors armed with carpet-sweepers and bin-bags. Lawyers acting for anonymous parties had brought pressure to bear through medical and judicial channels in order to procure—or at least secure—any files pertaining to their clients. It was a murky business, full of counter-claims and jurisdictional wrangling, and with no resolution imminently pending, it had been ruled that the files must remain in a secure repository so that nobody—in particular the police—could have a thumb through them in the meantime.

The apparent flaw in this state of affairs was, of course, that the evidence repository was a police facility, but this was not something that particularly disquieted the lawyers. Angelique had first-hand experience of the place from two different operations that had gone down in Marseille, enough to understand that the red tape was harder to get past than the armed guards. Everything needed paperwork and every movement was logged, including who got admitted, what items they were granted access to, the works. Of course, a cop connected or desperate enough could call in a few markers or grease a few palms to get the right person to look the wrong way, but if your later prosecution hinged on anything the defence could demonstrate as sourced to the forbidden files, then your bad guy walked. Hence the bent lawyers' sanguine attitude to the arrangement, and hence it was in the cops' own best interests to make sure nobody—especially nobody in blue—got access to those files.

That was why Angelique had no feasible resort to official channels, regardless of her connections, but then she wasn't looking for a conviction: just anything that would let her get the drop on Darcourt before anyone else.

She looks across at the repository compound, a couple of hundred yards away on the other side of the boulevard. She can't see the two armed cops on patrol inside the perimeter, but she knows they're there, and that they—or their relief—will be there all day and all night. Nothing's changed since she was last here: same reinforced concrete walls, same barbed-wire rims, same brutalist low-rise bunker housing the goods, same power-locked electric gate. She's even pretty sure she recognises the guard on duty at the gatehouse, squirrelly little grey goblin of a man with an exasperatingly fastidious attention to bureaucratic detail. She remembers once musing to herself whether he was armed principally to deter cops from shooting him, rather than anyone intent on raiding the repository.

This time, however, though she would have no gun with which to threaten him, and even worse, no paperwork, he was going to lead her right inside without asking any questions.

'You about ready to do this?' Zal asks.

'Make my debut as your sexy assistant? Sure thing.'

Zal puts the van in gear and drives it around the corner to the spot they selected earlier, the backlot of a closed-down restaurant. Zal parks it with all the precision and care of an SUV-owner in a cramped multistorey; ie slewed diagonally across three marked paces so that it looks abandoned.

'Okay,' he says, turning off the engine. 'Make the call.'

Angelique gets out of the van and dials a number that will get her through to a central police switchboard from anywhere in France. She asks them to patch her through to the local nick in Toulon, where she verifies her police status and informs them she's spotted a suspicious vehicle. Can't spare the time to look into it herself and don't want to tread on anybody's toes (always worth throwing that one in when it isn't costing you anything) but it looks abandoned and ought to be checked out.

The local officer thanks her for the tip and she hangs up. Meanwhile, Zal has opened the rear doors and offers her a considerate but unnecessary hand to climb inside. She takes position, like he's shown her. She looks up to see him scrutinising, but he's not merely making sure she's doing everything right. Something seems to strike him. Zal can always hide what he already knows needs to be concealed, but whatever this is, it came from nowhere and managed to play on his face for a moment before the curtains came down.

'What?' she asks. 'Everything okay?'

'Just thinking you better not sue me if you get deep-vein thrombosis.'

Nah, Zal, that wasn't it.

'Get going,' she says, a smile her way of concealing from him that she knows he's lying.

*   *   *

Zal closes the doors and walks swiftly away from the lot towards where their second, locally hired car is parked. He hardly expects they'll come flying round the next corner with their blue lights flashing, but he wants to be out of sight quickly nonetheless. He hid it from Angelique, but he's twitchy, disproportionately so. It could be because he hasn't pulled anything illegal for such a long time, but who's he trying to kid? It's because it's her, and she's on her own now. It's his plan, and it's now out of his control, but what's worse is, he hasn't given her an alternative out if it goes wrong. Ordinarily, that's a cardinal sin, but she insisted she wanted to go ahead, arguing that if it came to it, she did have a genuine 'Get out of jail free' card.

He knew he'd have to mask his anxiety when she lay down in place, as he couldn't let her see him look worried when *she* was the one about to be left on the line. Maybe that's why he got blindsided by something else as he watched the grace and speed with which she climbed into position. It was a

361

fleeting vision, a glimpse of a possibility that could change the whole . . . No. It was a dumb fantasy that he'd be embarrassed to recall, or just another way of pulling himself apart for nothing.

His phone rings about forty minutes later. Jeez, he hopes he never gets mugged in this town, if that's the cops' response time when you hand them something on a plate.

'Monsieur Innez?' the cop says, same Anglophone from this morning. 'We have good news. Our officers have found your van.'

'Is everything all right?'

'A window is broken, and it has been, how do you say, "hotwired", so there is some damage to the steering column—'

'Forget the van, is my property intact? That trunk is invaluable to me, I cannot stress this enough.'

At this, the cop pauses a moment, the bad news beat. 'The box is still inside, but I must inform you: my officers have reported that they had a look inside, and unfortunately it is empty.'

Zal lets out a laugh of precisely measured relief. 'That's okay, it was empty before. It's the trunk itself that I'm concerned about.'

The cop sounds twice as relieved as Zal pretended to be, at the sound of a major headache just vanishing from his case-book.

'In that case, it sounds like we could have a happy ending. If you still have the vehicle's keys, perhaps it would be best if you met the officers where the van has been found, and then you can examine both the vehicle and the—'

'I'm in Marseille,' Zal interrupts, sounding suddenly anxious and insistent again.

'Yes, sorry. You told me this morning, you have
. . .'

'That's right, a very important engagement which I can't cancel, though I wish I could. I'd drive to you right now, but . . . goddamn it. I'm sorry, I'm just so paranoid now, I mean if something else happened, after . . . Look, is there somewhere safe—I mean really, really secure—where you can store the trunk until I get back?'

*　　　*　　　*

Angelique holds off until she starts to fear her left bum-cheek is beginning to atrophy, the luminous dial of her wristwatch assuring her that the repository is long closed and well into its overnight regimen. Her fingers trace along the inside edge until they find the stubby release lever, operating it as Zal showed her, the double-catch mechanism designed to prevent it being triggered by accident. She flips up the hinged false bottom, then opens the lid the most delicate sliver to look out. She can see nothing. The vault is in darkness. Good. She slides a slim but powerful torch from her sleeve and uses it to scope her surroundings. The fine beam picks out an array of ugly adjustable aluminium shelving units, bearing the most randomly eclectic cornucopia of improbably juxtaposed objects this side of the Turner Prize. Not so much Aladdin's Cave, more Ali Baba and the Forty Thieves, or at least where the Arabian polis stored the Forty Thieves' haul while they did Ali Baba for reset.

Picking out and confirming that the only door (a steel, windowless affair, just as she recalled) is

363

closed, Angelique climbs fully out of the trunk and finds herself among the heaving shelves of the repository's claustrophobically cluttered—but methodically labelled and catalogued secure— vault. As she had confidently predicted, the local cops didn't want something as bulky as the trunk clogging up their already cramped and busy station, and had opted for the greater space and higher security of the repository, which also happened to be conveniently close to where the goods had been found anyway.

They checked inside it when they first got to the van, which precipitated a discussion between the two investigating gendarmes as to what had been inside it and therefore whether both trunk and van would need to remain in situ and be dusted for prints. Shortly after, however, it was relayed to them that there was nothing missing and they got their orders to shift it just around the corner. There was a short wait for another vehicle, presumably one bigger than whatever they'd rolled up in, before it was transferred amid a reassuring minimum of huffing and puffing, and driven the short distance to the repository. Upon arrival, it was opened for inspection again, presumably by Le Goblin, before being deposited directly at Angelique's intended destination.

Her only real worry had been the weight, but even that, Zal assured her, was accommodated within the illusion. Despite being made of light (though sturdy) materials, the trunk looked like it ought to be heavy. It looked old, it looked finely carved and decorated, and most importantly, it looked expensive, which lent expectation of heft to any object.

She had no such concern about there being any visual clue to the false bottom. When Zal first opened the trunk and let her see inside, she thought he was kidding, presumably having the real one ready to roll out once he'd had his fun. If there was a hidden compartment at the bottom of that box, then it couldn't be any deeper than about three inches. She was petite and she was supple, but she wasn't a fucking jellyfish.

The secret, though, was in the skirts. As the ill-fated Aberdeen manager Ebbe Skovdahl once said of the mini variety, comparing them to statistics, they suggest much, but they hide more than they reveal. The woodwork around the bottom of the trunk was ingeniously constructed to suggest that the underside of the box sat higher than it really did. The illusion was further enhanced by the paintwork on the exterior, which confused the eye into perceiving the box shorter in height; and by even subtler paintwork tricks on the inside, which suggested upon a cursory glance that the outside edges of the bottom were in fact the bottom-most inches of the interior walls.

She enjoys a near-orgasmic stretch, arching her back and extending every last sinew to work out the knots tied by her confinement. Then she gets to work.

It's a good thing she's got all night. It takes long enough to locate the Bouviere files with only this tiny torch to search through the gloom, but that's just the beginning of her quest. Bouviere's practice had been running for more than two decades. There are hundreds of files to flick through, and nothing to immediately identify the ones she's after, it striking her as just a little too much to

365

hope that Darcourt would have presented under his own name. Given the nature of the surgeon's speciality, an accompanying mugshot isn't going to jump out at her upon a cursory glance either, unless there's a 'before' picture included for reference.

She has to pore over each file methodically, finding quick filters for elimination where appropriate, such as sex, ethnicity, age and extremes of height and weight. Beyond that, she has recourse only to her own memory of what Darcourt looked like, and to the printout of the police artist's impression she had them email to her that morning. She got the sketch—and the latest—from DC Ishtar Mitra, Dale being unavailable.

There had been no further developments since Darcourt released Sally Smith. The poor girl turned up on a tube train at Leicester Square, her distress and disorientation exacerbated by mobbing scenes reminiscent of a Beatles concert, before she was finally rescued by two officers of the London Transport Police. She was recovering in hospital but had so far been unable to tell the police anything unknown to anybody with a web browser. She remembered nothing about her abduction, other than feeling sleepy inside the limo. She never saw her captor, she said: not she never saw his face—she never saw him at all. She woke up in her cell after passing out in the limo, and stayed there for the duration of her ordeal before passing out again and waking up in central London. She had no information about Anika's fate, said the cells were soundproofed so they couldn't talk to one another. She knew Wilson was dead, because Darcourt told her, but there had been no such announcement regarding the third member of the group.

She was at least able to tell them that the t-shirt she was dressed in when she was released was not her own, and thus confirm that the slogan it bore—'There cannot be sin'—must be a message from Darcourt. Typically pompous, pretentious and self-indulgently cryptic, Angelique thought. Fortunately, the media hadn't cottoned on to it—yet. According to Ishtar, Sally had her hands up over her face in fear during the initial camera-phone onslaught, her elbows obscuring the lettering on her chest, and in the subsequent paparazzi shots taken coming out of the tube station, she had a blanket draped around her by her police escorts.

It was the end of her ordeal, but it had just been the beginning of the siege. Cops getting in and out of the hospital where she was taken had to negotiate a gauntlet not only of news media, but of the several hundred fans/rubberneckers who were keeping permanent vigil outside.

Ishtar asked for Angelique's tuppence-worth on the t-shirt. She told her it was most likely another red herring to waste police time and get the public speculating. Nonetheless, she passed on Ray Ash's phone number anyway, telling Ishtar that if it was of any genuine personal significance to Darcourt, then he was the man who would know.

If they were keeping the Great British Public in the dark about the slogan, Angelique hoped they were doing the same with the Baker sketch too. Based on a frightened man's recollection of five-year-distant glimpses, it bore almost no resemblance to the Darcourt of Angelique's memory, or of any existing images of the man. Unless Bouviere had carried out the greatest max-fax job of all time in altering his appearance beyond recognition, she couldn't see how the sketch was going to help anybody, least of all her in this particular task.

In the end, it's not an image, but a name that stops her riffling fingers and causes her to catch her breath.

Matlock, L.

Its robustly percussive consonants stand out amid the mellifluence of so many southern European surnames, and something about the word itself triggers an association.

She opens the file. The L stands for Lydon. Lydon Matlock. A self-indulgent Darcourt alias if ever she saw one. He had assigned both monikers as codenames at Dubh Ardrain, using the names of the remaining Sex Pistols and two other bands to address the rest of his team.

It seemed even the needs of secrecy and discretion were not entirely paramount when Simon's ego needed nourishment.

She places the file down on the floor and plays the torch across its pages, spreading them out and immediately scanning the ones bearing art. Bouviere carried out extensive reconstructive work in September 2001, some of it restorative due to severe facial abrasions and other injuries. Seems poor Simony-wimony got a wee scrapey-poo and needed a sticking plaster or two, and got the surgeon to give him a whole new face while he was at it. It was a complex, staggered procedure. There were several operations, each detailed in print and accompanied in some cases by X-rays and printouts of computerised models. It's impossible to know whether they show intended effects or intermediate stages of the process, just as the few close-up photographs of bruised flesh and open tissue could be from the operations or merely documents of Darcourt's wounds. But having spied a few flakes in the stream, Angelique then discovers there's gold in them thar hills. Clipped to the back of a previously overlooked page, she finds a black-and-white ten-by-eight of Darcourt— under anaesthetic, hence he never knew it was being taken—that is as good an 'after' facial shot as she could have hoped for. He looks almost as distinct from her memory as the Baker sketch, and the closed eyes don't help, but it's definitely him.

Having uncovered one such nugget, she decides to mine deeper. She begins more carefully reading each page of the notes, starting with the most recent, the mere date of which causes her to double-take, as it is only a year old, and bears the referral details of a major hospital.

That's when she hits the true motherlode.

And now nobody's going to have to call Raymond Ash. The message is suddenly very obvious, in light of what is now before her eyes in the file of Lydon Matlock.

'There cannot be sin,' the t-shirt said.

When there's no future.

*   *   *

Zal's displays of impatient anxiety at the police station, followed by effusive, relieved gratitude at the evidence repository, require very little acting.

369

Presenting himself as an individual consumed by the need to be restored to what is most precious to him is something of a Stanislavskian performance.

He spent a long night alone in a motel room. He knew it must have been pretty long for Angelique too, but at least she had a task to distract her from her worries. All Zal had to do was wait, and he found it a tough shift, with fretting that she was okay being only a part of it.

Zal was used to being alone, sleeping alone, floating on an isolated island of himself as much as he was separated from the world aboard the *Spirit of Athene*. Angelique had only been back in his life a couple of days—and hardly a carefree and pleasant couple of days—but already her absence was an ache, like he was more used to the feeling of being with her than the years without her had ingrained.

He told himself it was just an exaggerated feeling, symptomatic of his worry: worry that she was safe in the mechanism of *his* scheme, worry about what he was getting himself into, worry about what dangers they might both yet face. And while all of this was valid, it wasn't what was most clawing at him.

The truly disquieting thing, as he lay awake in the dark, was that he found himself starting to wonder if there perhaps was a way to be with her. It was a dangerous thought, a virus that had to be contained because it threatened to undermine his resolve and lay them both vulnerable. He had to concentrate on keeping them both alive, and in the event that he pulled that off, the only further consideration he could afford was in steeling himself to once more let her go. The effect of this virus was that every hour they spent together threatened to weaken that steel, and could only end in them both getting hurt.

It was dangerously distracting, and furthermore, it was self-deceiving too. The ideas he unconsciously set in motion before he could act to derail them called for things he could not ask her to give, sacrifices he *would not* ask her to make; not least because he could not put her through the awkwardness of her inevitable refusal, which would be made all the harder for her by any sense of debt or obligation his present actions might engender.

They had been together again a couple of days. It was hardly a laugh riot, with Angelique understandably withdrawn into her own worries, but she seemed content to listen to him talking about himself. He had yammered away, often prompted by her specific queries, about how he found his new career, about Morrit and Lizzie, about his dad's professional legacy, about how he evaded Fleet, and about the offers beckoning him to move on. However, she didn't ask him why he never showed up at the Musée d'Orsay. It didn't take him long to deduce that this was due to her not knowing he hadn't—*because she never showed up either*.

Zal made a life of fooling people, but he couldn't fool himself. The illusionist has to have the most disciplined respect for immutable reality; nonetheless it didn't stop it hurting, couldn't stop him dreaming.

He has to restrain himself from kissing her once she has freed herself from the trunk, in the process waving his offered assistance away as quite unnecessary.

There has been no reprise of what happened in the dressing room, and no discussion of it either. It's not so much that they are pretending it didn't happen, more a mutual acknowledgement that it's something they can't afford to deal with right now.

Angelique is clutching a manila folder, a look of ominous concern on her face sufficient to suggest the file has a portent she is unsure how to express. Either that or the goddamn thing was useless.

'You get what you were hoping for?' Zal asks. 'Is there a headshot?'

'There's a post-surgery photo, yes, but that's not the only artwork. I think this file also contains the bigger picture.'

She hands him a single page from the folder, pointing to an official stamp stating *Department d'Oncologie*.

'You'll have to help me with my French,' Zal tells her. 'What does it say?'

Angelique swallows, clears her throat.

'It says Simon Darcourt is dying.'

III
The Perfume of Heroic Deeds

LIVING WILL

So how do you like *them* fucking apples?

All those jobs, all those hits, all those years, and I never once saw a death warrant in the form of a dossier in a manila folder like you see in the movies. Yet that was my cue to commence my current project: an A4-size file, containing every detail I needed, complete with ten-by-eight photographs. Small scale, really: single subject, just one individual who had to die, and it would be hard to find anyone who'd argue that this fucker didn't have it coming.

Get it now?

Lymphoma.   Haematological   malignancy. Inoperable. Inexorable.

Lots of harsh little words, not much in the way of wiggle-room or interpretation. Here's two more: twelve months. Have to laugh at that, don't you? The standard figure from the classic 'doctor, doctor' joke, or the set-up for a thousand hypothetical questions.

'You have twelve months to live. *Comme ci, comme ça*. Maybe more, maybe less.'

Can I have a second opinion?

'Yes. You're a cunt as well.'

Can't argue with that diagnosis. Let me be honest enough to admit that it would be, to say the least, inappropriate to whine about the injustice of it all. I'm not expecting any sympathy cards or floral bouquets, no messages of support from well-wishers who 'know I'm a fighter'. However, it does change people's attitudes, quite unavoidably. Once they find out about this, they'll know I'm racing for pinks. In fact, it's very hard to imagine a more compelling motivator to get your shit together and make one final, telling contribution, to carve out an indelible legacy for the world to remember you by.

The time has come to put my affairs in order.

*   *   *

Amazing what death, or even the imminent threat of it, can do to your reputation, never mind your profile. It put Four Play back on the top of the charts, for a start. Best career move they ever made, getting into that limo. A number-one single, a new, repackaged Greatest Hits album complete with said new 'classic' rescued from its filler-track purgatory and given the profound status of valedictory closing number, plus nothing but fond memories in the collective unconscious. Damaged septums excised from the picture more thoroughly than any cosmetic surgeon's scalpel could have managed; hookers, infidelities, cat-fights, contested paternities and date-rape accusations forgiven and forgotten.

Isn't it grand, boys, to be bloody well dead?

Can't see it working for me, unfortunately, though I'd like to think a few folk might rise above the hypocrisy enough to raise a private glass in recognition of one or two of my achievements of which they secretly approved. Not that it will be bothering me where I'm headed.

But you don't even need to have snuffed it for the eulogising to begin. Look how much love is coming the way of poor wee Anika since Sally

376

turned up alive and deductive assumptions were made about second prize. Assumption number one is that it was Anika who *got* second prize. Doesn't look that way to me, though. Sally is a star now, unimaginably more famous than her silly dance routines, pouty posturing and bratty precociousness could ever have made her, but I've turned Anika into a goddess.

Anika's past indiscretions have been effectively expunged from public record, and the two-faced, back-stabbing ruthlessness with which she secured her berth in Vogue 2.2 transmuted into 'drive', 'ambition', 'spark' and even 'a go-getting lust for success that we would do well to instil in this lacklustre generation of unmotivated youth'. But it's not just that her past can no longer be used to taint her: nor can her future. She's an ideal, an icon for people to project on to without fear of the reality spoiling the effect. Everybody loves her now, because there's nothing she can do to get on anybody's tits. She can't run off at the mouth and say something that will alienate anybody who disagrees with it. She can't betray her semi-literate ignorance of the written word as it exists beyond *Now* magazine. She can't let slip some racial or homophobic epithet that would bring down instant condemnation or call into question the values and attitudes she was raised with.

She's Princess Diana. She's Grace Kelly. She's Marilyn Monroe. No, she's greater than all of them, because unlike her goddess predecessors, her worshippers are about to be given the chance to bring her back.

Much as it would ordinarily make me sick, this eulogising was in fact music to my ears. With every word of it that was spoken or printed, she became an ever more precious commodity when, yesterday, I revealed that she was still alive. Alive and looking well, in fact, compared to recently, and certainly compared to Sally, who seems to be finding the spotlight a little harsh since she discovered there's no way to turn it off.

Anika's been moved to a new room with *en suite* facilities, toiletries, fresh clothes, a microwave and a fridge full of food. There's even—tantalisingly—a telly, which precipitated a real-life *Wacky Races* rally event as competing outside-broadcast vehicles rushed to reach her family and friends to relay messages of love and support. Sky News scooped the grand prix with Mummy and Baby Sis, ITN the silver with estranged Dad, and shame on the slowcoach Beeb, who trundled home with the consolation prize of Nell Devereaux, Vogue 2.2's publicist, which was a bit like getting a wank from a nymphomaniac.

It was all terribly fucking poignant, wasn't it? I was almost filling up myself once or twice. Heart-rending stuff, especially the way everyone's initial relief gave way to the horrible understanding that her ordeal wasn't over. Yes, thank God, she was still alive, but the question remained, as the Sky reporter underlined live from outside Mummy's Bromley semi: 'What sick game might Simon Darcourt make her play next?'

What indeed.

Well, I'm not going to tell you just yet, because I'm still in the business of assembling her new playmates: an all-celebrity line-up.

Our first stop is in a luxuriantly verdant and paparazzi-deterringly secluded enclave near

Cheltenham. As we pull up at the electronic security gates of a substantial mock-Tudor villa set in several acres of landscaped gardens, and bordered on all sides by privacy-ensuring arbour, allow me to put on my best Loyd Grossman accent and ask: Who lives here?

The answer is none other than Chelsea centre-half and England captain, Gary 'Nails' Nailor: winner of thirty-eight caps, two Player of the Year awards, two FA Cups and two Championships, but more significantly, possessor of one trophy that has ensured more column inches than all that silverware combined, in the shape of WAG almighty, Charlotte Westwood.

Nails might have commanded an eighteen-million-pound transfer from boyhood heroes West Ham and an image-rights deal that has his face earning serious coin both for and from Pepsi, Nike, Vodafone, Gillette and Sony, to name only five, but the boy's a fucking hermit compared to his missus. 'Champagne' Charlotte these days, when she's not guesting on any chat show, celebrity cook programme, celebrity dance programme, celebrity property programme, celebrity quiz programme, celebrity auction programme and celebrity relationship-advice programme that will have her, and when she is not being snapped for magazines coming in and out of upmarket eateries, Knightsbridge designer boutiques, West End film premieres and launch parties for products she officially endorses, can be found in the other pages of said magazines and in the breaks between the above TV shows, advertising everything from her own perfume, toiletry range and poseable dolls, to hair products, make-up, yoghurt drinks, chocolate and women's razors: electric and conventional. She has an agony column in one tabloid, a picture-led diary in the colour supplement of the Sunday edition of another, a ghostwritten autobiography in the top ten and a workout DVD launching in the summer.

Not bad for a finishing-school dropout with a single O-level to her name (in Arithmetic, no doubt handy for counting the money). Didn't hurt that she's a looker, whose trust fund bought her a flash enough wardrobe to get her into the right parties, where she met the emergent young football heart-throb with whom she was to embark on a fairytale romance that was the stuff tabloid dreams are made of.

I'm not talking about Nails, by the way, though you'd be forgiven if you've forgotten. No, in her bid for the celebrity stratosphere, the first booster rocket to Champagne Charlotte's space shuttle was Keiran Kelly, the nineteen-year-old Bolton Wanderers winger whose flowing blond tresses were about all most defenders saw as he skinned them on the way to the by-line. The only people coveting him more than a million dreaming teen princesses were the managers (football *almost* as much as marketing) who wanted him on their clubs' books. His twelve-million-pound move to Man United consecrated his and Charlotte's places among the elite of their respective fields, but Charlotte's long-term prospects turned out to be less susceptible to ligament damage. 'Special K', as he used to be known, had his career effectively wrecked by a late tackle against Middlesbrough; so late, in fact, that Keiran would have sworn he was still in a relationship with Charlotte Westwood when he took the ball past the full-back.

When the doctors announced that he wouldn't be able to play the game to professional level again, she shouldered the responsibilities that any WAG worth her salt knew were imperative upon her: ie she fucked off and found the next rising star—those responsibilities being to her own career and to an expectant media. Played it very cute too, making herself look like the one who deserved sympathy even as she ditched this poor fucker in his hour of greatest need, by leaking a slew of 'My secret hell with Keiran the love-cheat' stories to herald the break-up: material she must have been sitting on for use in emergency. She knew that, with the exception of the paper to which Keiran sold his side of the story, the media would spin it her way, because she had a big future selling their newspapers, and Keiran had a future selling insurance. She came out of it not only smelling of roses, but more high-profile and alluringly eligible than ever, perfectly placed to browse the form book and take her pick of the field. She put her money on a more robust thoroughbred, trained at Upton Park but recently stabled at Stamford Bridge, and it turned out he was a performer over long distance. Consequently, she has become the most ubiquitously visible female in the country: inestimably wealthy, indiscreetly lusted after, covetously aspired towards, inexplicably (but quite indisputably) popular and, this morning, expecting me.

I know. How *ever* did I swing it? What you wouldn't give etc etc.

Well, I called ahead. Makes all the difference. Called her personal publicist, anyway. It's true

Charlotte's a busy and important lady, but for all of that, she is still happy to give her time for an appeal, particularly if it is an appeal to her insatiable appetite for maximum-exposure, guaranteed-positive publicity.

'I realise this is very short notice,' I said to her publicist, a conspicuously trying-too-hard female going by the ostentatiously absurd name of Fizzy Brill (real name Philippa Brylle-Havilland; Papa led the Hestbury Hunt until the animal rights Khmer Rouge got their way). I gave my name as Gavin Aldlake, from a production company called Azimuth; told her we mostly made arts documentaries for BBC Four, which ensured neither she nor Charlotte would be familiar enough with the field as to wonder why they'd never heard of us. 'But I think you'll understand once I explain the circumstances. We've been subcontracted to put together a package of support and solidarity messages for Anika, to go out across all channels. I don't know if you've heard [yeah, right], but she's now got a television wherever she's being—'

'Yes, absolutely.'

I thought for half a second that she was merely confirming she knew about Anika's smart new Sony, but in fact this was her already agreeing to the proposition. I reckon it's fair to say it thus: I had her at 'all channels'.

'See, we want Anika to know that everybody's thinking of her, especially other celebrities, and Charlotte's, if she's happy to do it, would be the first message.'

'Oh, she'll do it,' Fizzy asserted with absolute confidence, before adding, 'I mean, Charlotte's

382

been very disturbed by all of this. I think any way she could help out would mean a lot to her.'

'I really appreciate this. And I really do stress, the word here is solidarity, not just support or thoughts or good wishes. This is about celebrities showing that they won't be bowed by what's been going on, that they won't be made afraid to be famous.'

'Oh, absolutely. I mean, Charlotte and I were talking about this recently: this Darko guy, it's like he's attacking celebrity itself.'

'Well, the fightback starts here.'

'You bet. I'll just check Charlotte's schedule for tomorrow. She *is* planning to be in London anyway; are you in Soho?'

'Oh, no, not at all. We'll come to her. This is a ten-minute job, two lines to camera. She can do it literally on her way out of the house.'

'Even better. When can we expect you?'

I slide down the window on the limo and speak into the intercom. A few seconds later, the electronic gate withdraws behind the wall and I drive on, stopping just inside. The gate has an anti-crush sensor, so a piece of tape saves me the slightly messier job of cutting the power line. There is a small CCTV camera pointed at the car; it's unlikely to be connected to recording equipment, but I've got my peaked cap angled down and my shades on anyway.

They're waiting for me on the front steps as my limo crunches the ochre chips: Fizzy, the good lady herself and a third female, who turns out to be her hair-and-make-up artist, called in at this early hour because it's for telly. There's also a bloke in sweatpants and a t-shirt whose status remains

383

ambiguous: he's either her minder or her personal trainer. He doesn't do or say anything anyway, just hovers around in the background folding his arms in a manner conspicuously intended to emphasise his musculature.

Charlotte and Fizzy both look a little confused when it's only the driver who emerges from the limo, but are instantly reassured by the sight of a video-camera and tripod under my arm.

'Yeah, it's just me. I'm Gavin,' I explain, briefly shaking a few hands amid the operation of setting up the tripod. 'We're stretched a little thin, what with some of the celebrities we've asked being fairly scattered about the country. My assistant, Scot, has been dispatched all the way to Inverness because that's where Gordon Ramsay is today.'

They like this. Gordon Ramsay being in the package is a favourable endorsement of the production's celeb calibre, but the fact that someone more junior has been assigned to him goes down even better.

'The thing is, it's aesthetically ideal to be doing it this way, because we're going for a kind of low-fi, docu-style verisimilitude feel. Makes it seem more immediate, more real.'

Charlotte nods approval, though I'm guessing it's the first time she's heard the word 'verisimilitude'.

'All I need is essentially two shots. One of you just walking up and climbing into the back seat of this limo. Then an interior; as you can see, there's a camera already set up on the dashboard. That's why I'm in this outfit. I won't be directly in shot, just on the periphery. Don't know if Fizzy explained, but—'

'Yeah, she did,' Charlotte confirms. 'Bit spooky, but that just underlines the point, doesn't it?'

'Exactly.' And people said she wasn't the quickest.

I make her do a couple of takes on the walk-up shot so that I look professional and nobody gets suspicious. She insists on a third, making out she's not happy about her own 'performance', but more accurately not happy about the gust of wind that blew a strand of her hair into the corner of her mouth just before she climbed into the back seat.

'You happy with that one, Charlotte?' Fizzy asks, conveying to me that while I might be directing this, I should be aware that her boss is the one calling the shots.

Charlotte insists on a make-up retouch and some work on her hair before we do the interiors. I have to stop myself looking at my watch.

I wait until this bit of running maintenance is done and climb into the driver's seat, where I adjust the second camera mounted on the dashboard and pointing into the back. Meanwhile, Charlotte composes herself, and I really mean that her expression and posture are a work of studied composition. If she had any idea what verisimilitude meant, she'd still be saying bollocks to it: her job is to look good, whatever the circumstances. This piece of self-sculpture complete, I resist the temptation to ruin it by suggesting she be wearing a seatbelt in the shot (a better example to the kids than the late People's Princess and all that), and cue her for a dry run at her line.

I make a show of adjusting the camera again, shaking my head and tutting.

'Sorry,' I say. 'Need to get your entourage out of the picture. Won't be a sec.'

I climb out of the car and explain to the hangers-on that I need them to move around to the side of the house. They comply obediently, the make-up girl considerately gathering her bags from the top of the steps to remove them from view. I usher the trio around the building, out of sight of the limo, slipping my silenced Glock from inside my jacket and concealing it behind my back.

'Won't be long,' I assure them. 'Just three shots.'

Holstering my gun again, I return to the car and this time start the engine, which causes Charlotte to furrow her brow. This is as much surprise or curiosity as she will allow to register on her face for fear of undoing the above-mentioned composition.

'I reckon it will be more poignant if the car is moving—you know, that moment of pulling away towards unknown destiny.'

'Yeah, yeah. Got you. Absolutely.'

I lean across the passenger side and make a final adjustment to the camera. 'Your cue is when we pass that tree, so we can just make out that you're being taken away from the security of your own home.'

'Got you.'

I set the camera recording then put the car into gear.

'Okay, turning,' I tell her.

I drive slowly and smoothly away from the front of the house, Charlotte staring out of the side window with an infinitesimally disquieted expression that I assume is supposed to convey grace under pressure. As we approach her mark, she turns to face the camera and gives the subtlest shake of her head, briefly closing her eyes as she does so. It's supposed to express the depths of her pain at merely contemplating Anika's abduction, though would function just as appropriately in refusing a sprinkle of parmesan in an Italian bistro.

Then, as we pass the appointed tree, with her house shrinking behind her in the rear windscreen, she delivers her line.

'It could have been any of us, Anika. So we're not just thinking of you, we're *with* you. We're part of this. And we'll do anything it takes to get you back alive.'

I trigger the gas release in the rear and accelerate for the open gate.

'That's a wrap.'

*   *   *

It really turns into an exciting day for me, getting to meet so many famous faces. A lot of folk do get thrown a celebrity bone when they're circling the cancer drain, though, don't they? A wee bedside hospice visit or a photo-op for the local paper, but you're talking about a local celeb too, usually: someone ten years binned from *Coronation Street*, who's angling to publicise the panto they're co-starring in at the Fudbury Town Hall. Doesn't hurt the image either: you always look comparatively youthful, glamorous and coiffured next to some emaciated fucker on chemo. But most of all, it's a chance to give something back.

Lucky old tumour-ridden me, though, *three* in one day, and we're talking strictly household names here: nobody in immediate danger of having to say yes if the *Call My Bluff* booker happens to ring. There was Charlotte, first thing, of course, and next, just before lunch, there is Danny Jackson, stand-up comic and stalwart of 'light entertainment' in a career spanning four decades.

Yes, he's quite the survivor, our DJ. First graced the screens back in the Seventies among the pantheon of old-school (ie mother-in-law jokes and casual racism) comics rapid-firing feedline/punchline gags on ITV's legendary *The Comedians*. Legendary in as much as the myth of it has grown as the mists of time have clouded its memory: in reality it was a largely unregarded half-hour schedule-filler, slotted haphazardly in the early part of midweek evenings before the big shows that people really wanted to see. That said, it was the only such showcase for stand-up comics, and its regulars were therefore in their time the biggest stars in that particular firmament: Stan 'The Geeermans' Boardman, Frank 'That's a cracker' Carson, Jim 'Nick-nick' Davidson, Bernard 'Cholesterol? Never heard of it' Manning, and, of course, Danny 'I'm all right' Jackson. He was slightly younger than his peers, more photogenic and a great deal more image-conscious; this last an indicator that he was sharper than them, too, in ways they weren't even aware it was worth being sharp about.

While they were still doing the working-men's club circuit for their bread-and-butter, hawking the same act as they did on *The Comedians* plus swearing and jokes about periods, DJ was refining his cheerful Cockney cheeky-chappy front in order to transcend being a mere comic and become that most viable crossbreed strain: a personality. It was this evolution that kept him fronting game shows while his erstwhile peers became the entertainment world's living fossils, embittered relics of a bygone age.

When that revolting polyester suit full of sweaty corpulence Manning belatedly got the final hook in 2007, it was moaned that he and his ilk had been victims of political correctness, and that this nanny-state prudishness had stifled the careers of these talented entertainers. Danny Jackson was conspicuously absent from the tributes, because he knew he was living proof that this explanation was utter bollocks. Manning and co weren't victims of political correctness: they were victims of the same thing as had given them national fame in the first place: television. They were jobbing comics with a limited repertoire, which was fine when you were playing a different club every night and the audience was too pissed to remember you telling all the same jokes when you last visited eighteen months back. But television delivered your gags straight to a national audience, among whom it would be repeated and retold, in the workplace, in the pub and in the playground. Thus TV producers weren't refusing to book dinosaurs like Manning because their material was now considered offensive; they were refusing to book them because their material was known verbatim and they proved incapable of coming up with anything new.

Danny Jackson saw it coming, and not only pitched himself as a bubbly and loveable scamp just right to inject zany enthusiasm into otherwise predictable family formats, but boosted his stand-up career into the top rank by the cute tactic of keeping it very distinct from his TV self. It was a cleverly worked symbiosis, a masterclass in the art of having one's cake and eating it. Being a household name, on TV every Saturday teatime, meant it was never hard to shift tickets—or videos and later DVDs—for his live tours, but he never performed any of his stand-up material on TV. Well, he couldn't could he? It was saucy stuff, 'strictly adults only', as it said on the posters, perhaps because nobody in the audience would want their kids to see them laughing away at jokes about niggers, pakis, four-by-twos, jam-rags and shirt-lifters.

The symbiosis enshrined our nation's self-deluding hypocrisy. DJ's enduring success as a TV host was partly down to his popularity with white van man and his missus for his stand-up reputation and the values it pandered to. They identified with him because they 'could tell he knew where they were coming from', and it gave an edge to his appearances that there was this latent side of him that the viewer knew was being kept in check. The same nation that so hysterically and ostentatiously condemned football pundits and reality-show contestants for unguardedly letting slip isolated racial or other proscribed epithets, was nonetheless happy to have its Saturday-night family entertainment presented by a man who cheerfully indulged every kind of bigotry—just as long as he kept it off the box.

He was way too smart to ever break that particular fourth wall, and very careful to protect the gossamer-thin defence that his live material was merely a stage persona rather than his own opinions. Gossamer-thin because, as he once put it: 'It's like a condom. The thinner it is and the more real it feels, the bigger the thrill.'

Thus he was acutely conscious of the damage should he ever be caught riding bareback, as it were, which was why he once knocked back *Celebrity Big Brother*, and also why he said yes to my own request.

Clearly it wasn't the only reason he knocked back *Celebrity Big Brother*; there was no danger of the *Call My Bluff* booker even giving DJ's agent a bell, far less him having to say yes. Unlike the rest of the *CBB* contestants, he didn't need the money and he didn't need the exposure; but more than that, he couldn't *afford* the exposure. Though he wouldn't have known who else was going to be in the house, he knew how the producers worked. It was a certainty there would be housemates from the ethnic and sexual minorities, and he knew he couldn't be guarded and circumspect twenty-four/seven; nor could he rely on sympathetic editing to erase any little indiscretions, those being what the producers were principally interested in.

When the tightrope he walked was the width of that gossamer membrane, not only could he ill afford such risks, but he had to regularly assuage suspicion and innuendo through displays of public mateyness with a select few tame black and homosexual celebs. An upscale version of the 'some of my best friends' defence. That was why I knew he'd be more than happy to contribute a solidarity message for little Asian Anika. How could he say no? It reflected his A-list status that he was among the first to be asked, it showed he cared and was prepared to give up his time for a good cause, it demonstrated a deeper, less frivolous side to his personality, and it proved he

391

had nothing against the coons.

But I said three household names, didn't I? And I saved the biggest name until last; or at least our respective schedules dictated it that way: *AMTV* having her on-air, today as all days, until just before my meeting with the charming Mr Jackson.

I'm new to this whole TV business, but I like to think of myself as a fast learner, and I know that for a project such as this, the chemistry is crucial. I need balance, I need light to contrast shade, yin to counter yang. So as well as a spoilt narcissistic fucktard and an insufferably smarmy bigot, both of whom many people in my prospective audience would secretly enjoy seeing tortured and degraded, I need someone of whom nobody has a bad word to say, and to whom nobody could wish any torment; someone genuinely loved and admired throughout the country: punters, pundits and industry alike. And that's why, today, I get to meet the warm, the genuine, the lovely, the *wonderful* Katie Lorimer.

I know. Touch me.

The Katie Lorimer. A woman so infallibly adorable that if there was to be a Britain's Nicest Person title, all it would take was a single nomination for her and everyone else would voluntarily withdraw, rather than be so mean or arrogant as to even contest it. Everybody, *everybody* loves Katie.

It's perhaps because she seems so natural in front of the camera, so down-to-earth, like she just stepped through the screen from an ordinary living room and into the *AMTV* studio, the perfect everywoman, or everyhousewife at least. Which isn't to say she's anything less than professional, but that she doesn't seem yet another polished example of some metropolitan TV-presenter production line. She hasn't sprung fully formed from the loins of Zeus, or even sprung secretly from the loins of Hughie Green.

This is why she and *AMTV* are made for each other; it's hard to imagine the show being so popular without her, just as it's hard to imagine Katie's style on any other show. She's perfect for *AMTV*'s magazine format; not just because she talks to the guests like they dropped into her living room for morning coffee, but because her target audience find it reassuring that she discusses the day's news stories in a manner that gives the impression she doesn't understand them either.

All of which is to say nothing of her physical charms, crucial to extending her popularity across the sexes. She is no dolled-up glamourpuss, but for all that, has an approachability that crucially makes the ordinary bloke think he might be in with a chance. She's maturely mumsy without coming across as an imminently menopausal frump; which is not to say that she's plain, either. She's not overtly or provocatively sexy, but there's something of the stocking-top flashed beneath the housecoat about her. In short, she's the closest thing the UK has to a mother of the nation, but a mother the male viewer could plausibly imagine in a basque and sussies, taking it doggy-style with breathless alacrity once the nation has been safely tucked up in bed after a goodnight story.

I get to meet her on Ravenheath Moor, about ten miles from the village where she lives with the hubby and three kids. I sold her publicist the location on the concept of being driven off from somewhere mysterious, with sheer convenience helping seal the deal. And I'm happy to report that unless she is the ultimate pro and a flawless actress, then yes, she is everything people say, and that what you see truly is what you get.

She delivers the line as I pull away into open moorland, emerging from the copse of trees where her publicist, Elisha, and her driver, Tony, lie dead inside a black Audi A8.

'It could have been any of us, Anika. So we're not just thinking of you, we're *with* you. We're part of this. And we'll do anything it takes to get you back alive.'

Of the three I've filmed, she's the only one I would swear truly means it.

As I say, an exciting day all round, and an equally productive day's work, but not everything is in place for my farewell extravaganza quite yet. There's a slot open for one last star, and I'll be filling it very soon. Admittedly, this one is not a household name (though I'm about to change that), but it is a face that's been gracing all our TV screens recently, another admirable individual to balance that crucial chemistry I was talking about. Not a celebrity, but someone who's nonetheless been playing a public part in these shenanigans: my noble adversary from the long arm of the law.

This will be arguably my most valuable acquisition, as it will ensure that the cops think very carefully about all their options before attempting anything rash. Nothing concentrates their minds and stays their more gung-ho instincts quite like having a genuine stake in the game.

PREY

Darcourt is commencing his endgame, of that Angelique has little doubt. The haematologist she showed the file to confirmed as much. Lymphoma, leaving him functional but imminently doomed; just not imminently enough. If the bastard had just assumed the lump on his neck was a big plook, the late onset of other symptoms might have ensured he was too sick to engage in this valedictory undertaking. Nonetheless, his time *is* running out. The haematologist said, going by the dates on the file, he should be hitting the debilitating phase in a matter of weeks, if it hasn't begun already. That, in fact, might be what has heralded the commencement of what she is sure will be his final play, and though she yet has no idea what form it will take, she knows for certain that it signals her parents' time is fast running out.

The thought of them is like a low hum in her head, ever constant even if she can relegate it to the background some of the time. It is a unique aspect of this torture that she has to keep tuning them out: that the only way she can maintain the composure required to have any chance of helping them is by banishing them from her mind, or at least drowning her thoughts of them in a deluge of more immediate concerns.

Her most immediate concern right now, as she sits on the Piccadilly Line, is an unsettling, goosebump-pricking suspicion that she is being followed. It's not some indefinable sixth-sense nonsense, but a combination of trusted instinct and experience. She hasn't specifically seen anything to set off her sensors; rather, it is like a belated, subconscious awareness that something in the background may have been more significant than it registered at the time. Something glimpsed back on the platform at Holborn, perhaps, a furtive look or movement that some corner of her mind took note of while the larger part of it was busy contemplating the latest celebrity abductions.

However, she also knows that it could instead simply be down to the paranoia attending the purpose of her journey: her first clandestine meeting with Zal back on British soil, to ponder how best they might obstruct or at least illegally divert the course of justice.

As she rides the up escalator, she hears the chime announcing a text message, a sound that has come to elicit a Pavlovian quickening of her pulse and gut-churning, dry-throated dread. The woman in front of her dips into a handbag and produces a mobile, same model as Angelique's, same default alert. Angelique sighs quietly, her breath lost in the roaring of a train as it thunders away from one of the platforms below. That text wasn't for her, but there will be another winging her way soon enough, her puppetmasters continuing to remind her that they have seen the same straws in the wind, or at least the same news reports and images on the internet. The latest photo of her parents showed them holding a printout of Darcourt's homepage, just to underline the point. They looked ill. She tries not to stare at these attachments, these little digital cluster-bombs, wishes she could force herself not to view them at all, but the desperate thirst for news, for information, always overrides this. Each new shot shows them more anguished, more gaunt, more tired, thinner, and so much older.

It's not just pictures they send, though. There are usually questions, requests for updates. She answers to the best of her knowledge and tells them almost everything she knows, in case any of the queries are a test of her obedience and reliability. Their claim of having other sources in the police is not something she's about to call their bluff on, especially if they are 'sources who do not even know they are sources' and thus unaware what information they are betraying. If they decide she's no use to them, or that she can't be trusted, then it's over.

She has consciously suppressed only one thing, but it being the single most salient development to emerge from the investigation is the reason her heart jumps every time she hears that chime. Every new message could be the one telling her they've discovered she is holding out on them. This risk, though, she has ruled smaller than that of telling them that Darcourt has only months to live, as she has no inkling how this might change the plan their end.

The first herald of Darcourt having broken his blood fast had been the discovery by Gary Nailor, upon returning from training yesterday lunchtime, of the bodies of his wife's publicist, her personal trainer and her make-up artist, with Charlotte Westwood herself missing. A few hours later, Norfolk police found two more bodies in a car on the edge of Ravenheath Moor, the corpses turning out to be the personal publicist and the chauffeur of Katie Lorimer, also missing. All five victims had been shot once at close range through the centre of the forehead.

The new Darcourt website had been uploaded later the same day, in time for the late evening news programmes and the print media's first-edition deadlines. Meilis had known about it immediately, having set up a software script to trigger an alert whenever the site's structure was altered. Angelique, like just about everyone else still in the building, had rushed to the Operations Centre where the alpha geek routed the output on to the big monitors. The page displayed an embedded video beneath a constantly refreshing still frame from the existing Anika live feed (which clicking on the still would link you to), and a small panel, white digits out of black, displaying a one-hour countdown in progress.

The video showed Charlotte and Katie—as well as Britain's favourite family-friendly racist, Danny Jackson OBE—climbing into the same black limo, followed by a montage of contemplative interior shots as they were driven away. Then each was shown looking directly into the camera and speaking, Charlotte first: 'It could have been any of us, Anika.'

Followed by DJ: 'So we're not just thinking of you, we're *with* you. We're part of this. '

Then, lastly, Katie: 'And we'll do anything it takes to get you back alive.'

Within half an hour, a squad car dispatched to Jackson's house in Finchley found the place empty and a woman's body in the downstairs hall, executed the same way as the others. Within two further hours, the woman was identified as Carrie Kendall, a prohibitively pricey call-girl with an exclusive client list and previous for dealing cocaine, a substance subsequently located in abundance inside her shoulder bag.

When the countdown reached zero, the page began to animate itself. The self-updating grab from the Anika feed remained in place at the top, but the video panel multiplied itself until there were four identical copies of it, showing the final frozen image of Katie Lorimer's painfully concerned face.

All four images faded to black, then one by one, the first three changed to host self-updating grabs from three new feeds: three new rooms, three new prisoners. Finally, after a short delay, the last panel altered to show a fourth room, unoccupied, the image overlaid with a grey shadow in the shape of a question mark.

'Room for one more inside,' Dale muttered grimly.

There followed a few seconds of silence, as though the room—and perhaps even the country— was collectively taking a moment to fully register the enormity of what had just been revealed. Then it sounded like every phone in the building— landline and mobile—was ringing at once.

Angelique looked at Dale. 'What we gonna say?' she asked him, the 'we' part in truth merely a gesture of solidarity.

He sighed, the words 'let this cup pass from my lips' all but etched upon his weary expression.

'I don't suppose "move along, nothing to see here" is worth a try?' he had asked.

*   *   *

There's a knock at Zal's hotel room door, and he springs from the bed, turns off the TV, checks the mirror, inadvertently patting at his clothes. This is what she does to him: the moment she shows up, he feels like he's just walked onstage unprepared, and a magician *never* walks onstage unprepared. He thinks of a hundred nights he must have lain awake onboard the *Spirit of Athene*, wondering what if, and imagining a moment just like this, imagining Angelique de Xavia knocking on his hotel-room door. In the fantasies, it was fair to say, he was a lot cooler. They worked out far closer to what happened in the dressing room, whereas right now he's doing a passable reconstruction of his conduct in the hour preceding that, when he spotted her sitting by the edge of his stage and blew his trick.

He had flown in that morning, caught a flight from Malaga to Gatwick, sharing an airplane with two hundred people who had been sold some bad advice regarding UV-protection. Poor bastards had evidently spent the previous two weeks under the impression that ingesting alcohol for eighteen hours a day was an effective prophylaxis against sunburn. They really ought to sue somebody.

After Toulon, he had gone back to the ship to put a few matters in order, not least straightening things with Morrit. Prior to them hitting the evidence repository, Angelique had expressed her concerns about the timescale of what she was asking, not wishing him to drop everything just to end up sitting around indefinitely. He had therefore assured her that he could be in London inside twelve hours from anywhere on the ship's schedule, even the open sea, the cruise company

400

having a helicopter at their disposal for emergency transfers.

'If it makes you feel easier, I can work my ticket, see out my contract, but I'd be on permanent standby, ready to move whenever you pick up the phone.'

It seemed a mutual, unspoken understanding that such a mercy dash might never be required, and he had wondered darkly about that: how the weeks and months might pass, their communication fading away with Angelique's hopes. There might only be a call, or just a message, to say the worst, and that would be the end—of everything. All would be as it was, with him playing the ballroom each night, sometimes looking at the table where she sat and remembering a dream that died. But what she had found in Lydon Matlock's medical file meant that, for better or worse, her quest would end soon, and with a bang, not a whimper.

He decided he had to come clean to Morrit. He owed it to him: not just because he had so suddenly dropped everything and gone off at zero notice; not just because he had left the old man hanging on in recent months, waiting for Zal to make some kind of decision; and not just because he had kept so much of himself secret throughout the whole of their past. He owed him as a down-payment on what he still hoped would be their future; and by way of ensuring that he *had* a future, Zal had to come clean because he needed Morrit's professional advice. As he saw it, this whole game would come down to how they played a hostage exchange. Essentially this was the 'cup and balls' routine, literally the oldest trick in a very, very old book. It dated back thousands of years, documented in ancient Greece, Egypt and Sumeria before that. There was even a name in Latin for its practitioners: *acetabularii*. But the soul of magic is in finding a new way to perform an old trick.

Standing backstage among their props, Zal tells him everything: about his dad, about the Escobars, the bank, Angelique, right up to Toulon. He explains what he thinks he needs to pull off, shaping Morrit's advice according to what he has learned from another old-stager: that when it comes to any kind of heist, you plan the job backwards.

'The first thing you have to put in place is how you're getting out,' Zal explains.

'And I take it that goes for the girl too?'

'Getting her out? Of course it goes for . . .' he replies, before reading Morrit's scrutinising face and absorbing what he really means.

'I'll slip away quietly, suffice to say. Save her any awkward shit. She'll have enough to deal with . . . either way.'

Morrit is shaking his head. 'Should never end on a vanish, son. It leaves the audience uncomfortable. You have to take your bows.'

Zal sighs. He has just given Morrit a potted history of his relationship with Angelique; how could he expect the guy to understand in five minutes a conundrum Zal is still struggling with after five years?

'This audience won't be looking for an encore, Dan. She didn't come because she wants me. She came because she needs me.'

'And what do you want, lad?'

'What I want doesn't really play here. This isn't

about me, it's about her.'

'Bollocks. You're just setting up your out, clearing the obstacles so it's easier to leave. What is it you're running from? Are you afraid if you ask her, *she* might give up everything for *you*? Aye, that would really scare you, wouldn't it?'

'You don't know what you're talking about,' he offers sheepishly. Devastating comeback. Yeah, that would really put him off the scent.

'I know exactly what I'm talking about. It's called love, and you're arse-deep in it. Look at you: you're dropping everything for this girl, running off to put yourself in harm's way and asking nothing in return. Admirable, son, admirable, but ask yourself this: if she's worth all that, isn't she worth the wooing, too?'

'I told you, Dan: she doesn't want me, she needs me. Big difference. She only came because she's in trouble.'

'Aye, you say. But there's two folk get married every day in this world on account of a girl being in trouble.'

Zal opens the hotel-room door and Angelique steps inside, bringing a whiff of perfume and the smell of clothes that have been outdoors in the cool, more pronounced to an olfactory sense that has been holed up in an air-conditioned room for several hours. He closes the door and turns, expecting to see her back as she makes her way into the room, but instead finds her stepping into an embrace, which she holds, wordlessly, for ten, maybe twenty seconds. When at last she pulls back, it is to look into his eyes, and at this point, for some reason he speaks before it can turn into another kiss.

'It's good to see you.'

She nods, looking a little unsure of herself. 'Thanks for coming, Zal. For everything. In fact, I don't think "thanks" really covers . . .'

'Let's worry about that shit if I actually prove useful,' he interrupts.

She shrugs a little, and Zal goes on before she can say he's already been useful simply by holding her, or anything else that might confuse both of them as to what she really wants. 'Because it looks like quite a game we're getting into.' He grabs the remote and turns on the TV. It looks natural, logical, getting down to business, but really it's just easier than talking about the other stuff. The TV defaults to BBC1, showing a news report, the faces of the three new abductees inset in the top-right-hand corner.

'No kidding,' Angelique says, moving into professional mode also. 'He just scooped up three of the most famous people in the country, a major escalation from glorified talent-show contestants. There's going to be fifty million eyes on this, which is why I appreciate having a secret confederate, not to mention one with a gift for misdirection.'

'It's escalated further than that in the last half-hour,' Zal informs her. 'He's added sound, and he's letting them mingle.'

'He's what?'

Angelique stares at the TV, while Zal thumbs the remote until he finds Sky News, the channel he just switched off.

'Just happened while you were on your way over here. Waste of having big stars if they can't talk to—'

Angelique's phone rings before he can finish.

404

She holds up a hand by way of apology/explanation that she must take this call. He understands. It's everything that's impossible between them in a solitary gesture. He cuts off speaking, won't make even a sound that the person on the other end might ask about.

'I was on the tube,' she says. 'No signal. Yeah, I'm watching it now.'

A couple of news channels are streaming the feed, no doubt on a delay in case anything truly unbroadcastable suddenly transpires. His flip through the networks showed that all the news bulletins are carrying at least a bit of this footage, same as they all showed brief clips of hostage videos, but Zal thinks Sky News may be kicking the ass out of the public-interest justification. This isn't reportage: they're filling their schedule with a celebrity reality show, one infinitely more compelling and thus ratings-boosting than any predecessor, and it ain't costing them a cent.

The three new captives are shown in a central area, doors off it leading to their individual rooms. The inset feed is too small for the type to be legible, but the newsreader confirmed earlier that each door bears a nameplate, with one displaying merely a line of question marks. The room has, like the one holding the Anika kid, a microwave and a (larger) fridge, plus the addition of a sink and a stack of crockery. Other than that, the only items of furniture visible are two chaises longues and a widescreen TV mounted on one wall, upon which they can see a live feed of the original remaining hostage, who appears oblivious of their presence.

Angelique finishes her call and continues to

gaze at the screen, sighing gently as to suggest a controlled outlet of far greater rage.

'This won't stop at extended bulletins on Sky News,' she says. 'The networks will be carrying this live after the watershed.'

'After the what?'

'Don't ask. You'd have to be British, and even then it doesn't make sense. But they'll give him everything he could want: national prime-time broadcasting. The digital networks will be able to dedicate a whole channel to this, twenty-four/seven.'

'Didn't they do that before?'

'No. Not enough to see—just miserable human beings festering in cells—and nothing to hear.'

'Why weren't they miked last time?'

'Darcourt's sick joke—they got famous by miming, so he denied them a voice. He also denied himself an angle of interest, but he's learning as he goes. This part's new to him, but he's proving expert at the abduction bit. He took three high-profile individuals in one day and left us with nothing. Killed all the witnesses, left nobody alive to talk.'

'What about the limo, the Merc?'

'Sore point. We had high hopes for some kind of triangulation telling us roughly where he's operating from; we figured he had to have dropped each one off before hitting the next, unless he had them in the boot, which is unlikely given the number of hours they'd need to stay sedated, not to mention still breathing. We assumed he must therefore be based somewhere within a reasonable radius of London—like everything and everyone else in the fucking media. Find him on one camera, suss the plate, pick it up wherever else it surfaces on the system and trace his routes to a common source.'

'But . . .' Zal prompts.

'Exactly. Despite us having more CCTV cameras than anywhere else on the planet, turning the whole bloody country into a reality show, the only footage this Mercedes popped up on is in the clips of those three willingly climbing into it. It's like a ghost car.'

'Or maybe a Transformer. Have you put out an APB on giant killer robots spouting macho dialogue?'

'We're not that desperate yet, Zal, but give it time. We've bugger-all else. He was smart. Hit three in one day, got the last one in the bag before anyone had raised the alarm about the first. Though with every celeb in the country battening down the hatches, it's going to leave him a challenge filling that last slot.'

'Unless he's already got number four and wants to have a little fun with the speculation before revealing who it is.'

'That's certainly his style, though it would have to be someone nobody has noticed is missing.'

'You ask me, I'd rule nothing in or out regarding that question mark. From a showmanship point of view, that blank slot is worth more to him than the other three, A-list as they are, because right now it could be anybody: top of the bill, more famous than all of them. But speaking as a magician, what worries me about this is that it's a perfect means of misdirection. Everybody's focusing here, focusing on what's going to fill that window. Which ensures his next move will be something none of you sees coming.'

Angelique's phone rings again, and she gives Zal a strain-faced look by way of saying she must take the call. He gestures with a single open palm: don't sweat it. She speaks briefly then hangs up.

'Gotta cut it short, I'm afraid. Boss has had to call a press conference and he wants a wee huddle beforehand, see if between us all we can cobble together enough waffle to make it sound like we're anything other than caught with our pants down. Again. I'm sorry, Zal. Just got here, too.'

'Hey, I got a hug: big night for me. I've been assigned the lone wolf role on this one, and I'm cool with that. You go lie to the media, sell 'em back some of what they're usually shovellin'.'

'Okay. You work on this thing us polis won't see coming.'

He sits staring at the door for a long time after she's gone, the TV still burbling in the background but so distant from his attention as to be in the next room or even the next hotel. He's wondering whether it was his imagination or did she seem in a hurry to get out the door? Shit, leave it, man. She's in the eye of a storm here, not to mention being in the situation he chose to spare her all those years ago: of a police officer fraternising with a suspect wanted for armed robbery.

Her leaving felt odd, though. Incomplete. They had hugged in Toulon, at least, before saying goodbye. Tonight she just put her phone back in her pocket and walked. Was it because she had stolen time for a visit in the midst of what had to be frenzied police activities? Was it because they were now at any moment only a few minutes away from each other? Or was it that, despite their proximity, they were now moving apart by degrees? If so, then why did he, supposedly clearing obstacles from his path in preparing his out, fear that explanation so much?

A knock at the door seems to render it all moot. He looks round the room, can't see any belongings she forgot. He thinks: fuck the out. No matter how painful it gets, he'll deal with that shit later. Right now, let it be complicated, let it be her returning to say goodbye with a kiss, a hug, anything.

He opens the door and finds a gun stuck in his face instead. Okay, maybe not *that* complicated.

There's two men, middle-aged, just the chubby side of burly, both wearing grey suits, one balding but bullet-cropped and the other with a bad comb-over. They railroad him into the room, gun still pressed directly into his forehead by Comb-Over while Bullet-Head closes the door.

'Mr Innez,' Comb-Over says. 'Welcome back to Britain. Her Majesty would like to extend her hospitality.'

Bullet-Head turns around to face him, dangling a pair of handcuffs.

'For a minimum twelve years,' he adds.

*   *   *

Angelique can't think of anything that seems less important than this press conference as she leaves the Halton Court Hotel. It's not even window dressing, more a fig leaf of an exercise to cover the polis's collective nakedness. There's nothing to report: the public have access to as much information as the cops, so if they want to know the latest, they can log on or, it would now appear,

409

tune in. The only thing the police have an exclusive on is something they have no fucking intention of sharing right now, that being Darcourt's insufficiently impending demise.

She resents having to bail out on Zal, but truth is, she's just so jumpy about this massive act of deceit she's embarked upon that she feels she has to be super-keen little PC Shiny Buttons in order to allay suspicion. He only flew into town today, but she's felt like she's looking over her shoulder the whole time to see who might be watching. In fact, she finds herself literally looking over her shoulder as she walks away from the hotel. There were two men she'd have figured for cops walking towards her as she exited the building, but to her slight relief—if not complete comfort—they've gone.

It is only as she is descending into the tube station around the corner that her surroundings trigger a recall of whatever her mind had subconsciously flagged on her way here. Like jumping back three chapters on a DVD, she suddenly sees the ticket hall at Holborn again, except this time she knows what to look for. Grey suit, pot belly, bad Bobby Charlton effort up top. She had seen him at Holborn, then seen him again outside the Halton Court. It just hadn't clicked because there were—oh shit—two of them. They'd been walking towards her as she left, then disappeared: undoubtedly, she now realises, into Zal's hotel.

She turns and runs, full pelt, back to the Halton Court, where she storms up to the desk and sticks her badge in the receptionist's face like it's a revolver.

'Two men, grey suits, little hair. They just came in here. Where did they go?'

The girl looks utterly flustered and not a little confused, glancing from side to side like she's expecting someone else to answer for her.

'This is an emergency,' Angelique all but yells. 'Where did they go.'

'I think they . . .' she stumbles, long enough for her accented words to make Angelique worry she hasn't mastered the local tongue. She curses the fact that having learned several internationally popular languages, Polish has come up on the rails and overtaken most of them here in the UK. 'They said they were police too,' the girl eventually volunteers. 'They asked which guest you had been in to see.'

'Shit.'

Angelique is about to charge for the stairs, but stops and turns back to the receptionist.

'Did they show you badges?'

The girl looks apologetically helpless.

'I do not . . . I cannot remember.'

'Shit.'

'I am sorry.'

'Never mind. Just gimme a swipe card for room two-twelve, please. *Now.*'

The girl lifts a blank and slots it into the mag-strip dock, then begins tapping at the computer keyboard. Time passes. The girl stares at the screen, twitching a little, her cheeks hot with embarrassment.

'I am sorry, the system . . .'

Angelique remembers seeing a cart in Zal's corridor, a maid in the middle of her turn-down duties: an invaluable service for those folk

411

incapable of pulling their own sheets back or who can't make it to the pillow without the incentive of a square of chocolate to go that final yard.

'Forget it,' she says, and takes off.

She finds the cart one room along from where she last saw it, and flashes her badge at the maid.

'I need this swipe card,' she explains quietly, and lifts it from the tray at the side of the trolley.

Angelique stops outside a moment and listens. She can hear one man's voice, talking softly.

'. . . no, meek as a lamb. So we should have him hand-delivered within the . . .'

She swipes the card and steps inside. The guy with the Charlie Bobbleton is seated on the bed, his bullet-headed partner standing by the window at the far end, talking into his mobile. To her right, the bathroom door is open, and inside Zal is handcuffed to a towel rail. He raises his eyebrows in salute, looking sheepish to the point of embarrassment.

'I'll have to call you back,' says Bullet-Head, as Bobbleton gets up from the bed.

Angelique produces her badge and holds it up, by way of telling him to stay put.

'Police. What's the script, chaps? Let's see some ID.'

Bobbleton reaches inside his jacket with his right hand and produces a gun.

'Will this do, Officer de Xavia?' he asks.

Angelique takes her eyes off his for a brief glance at the weapon. Looks like an HK Mark 23; the vanilla rather than the laser-aiming model, but the main thing is it's not some Russian ex-military piece of shit, so it confirms these aren't gangsters. She clocked what they were one foot inside the

412

door: ex-cops, post-fifty-five, working private—and most likely corporate—security to fill the time and fill the coffers before retirement proper.

She knows there's no way this guy's pulling the trigger. She's a situation to be handled, an obstacle to get around.

'It's Detective Inspector, dickhead.'

'Oh, not once this particular chicken has come home to roost, you won't be. Don't know what games they played at your school, darlin', but the cops are supposed to *jail* the robbers, not fuck them.'

'I went to school in Glasgow, bawbag. Let me show you what we played.'

Approximately three seconds later, it's Angelique who's holding the gun, while Bobbleton is holding his face, lying on the carpet beneath her. She's sustained some damage herself: prick had a sap stashed in his sock and got her a sore one just above the eye before she put him down.

She was right about him not shooting, though: he had the safety on the whole time. There's no loyalty bonus going to make it worth his while to shoot a cop.

Bullet-Head remains where he was, observing developments with his arms folded and a bemused but calm look on his face.

'Who are you working for?' she asks.

'Client confidentiality forbids me from answering that question,' he says acidly, though there's a smugness about him betraying the fact that he'd just love to tell her if it wasn't more fun to dick her about.

'Gimme your phone, and the keys to the cuffs. Throw them both on the bed. NOW!'

He rolls his eyes and tosses the phone towards Angelique's side of the bed. As it hits the duvet, Zal emerges from the bathroom at her back.

Bullet-Head looks incredulously at Zal, then glares at Bobbleton.

'I thought you cuffed him.'

'Don't fucking look at me. "Oh, if we're going someplace, can I put my jacket on before you cuff me?"' he mimics, crap American accent. ' "Fine," you said.'

'Well, how was I to know—'

'The keys,' Angelique interrupts, lifting the mobile. 'Toss them, come on.'

Bullet-Head shakes his napper.

'It would appear your boyfriend doesn't need them.'

'No, but you and *your* boyfriend will when I cuff the pair of you to that same towel rail. Give.'

She toggles through the menu, looking for the outgoing-call log.

Bullet-Head folds his arms. She glances from the end of the gun to the screen of the phone. The last, truncated call was to 'D Holland'.

Bullet-Head is nodding with undisguised self-satisfaction.

'Get up, Arthur,' he tells Bobbleton. 'Officer de Xavia is going to stand aside and let us leave now, because Officer de Xavia really, really does not want to have to book us in at the station and explain any of this, now does she?'

*   *   *

Angelique is present at the press conference in body only. She stands a few feet to the side of the backboards, out of the television cameras' sightlines, her eyes on Dale and Aldwyn Keen. Kudos to the commander for sharing the stocks, the brass usually only making themselves available when there's credit to be doled out, but nothing he or Dale says registers in her head, and not merely because it's worthless.

It's a good thing she wasn't asked to help front this latest excuse-and-apology showcase, as she can feel a swelling fast beginning to grow above her right eye. She managed to stop the bleeding before getting to HQ, but despite buying a bag of frozen peas at a Tesco Metro and applying it throughout the walk here, that prick with the lead sap has made his mark. It was noticed, too, Dale immediately asking what had happened. She said she walked into the back of someone carrying an easel on the tube. It sounded embarrassingly daft enough for it to seem genuine, but her paranoia imagined him divining all sorts of revelations from it.

Debbie Holland, RSGN's corporate zombie bitch: that's who had sent the male-pattern-baldness brothers after Zal, and what made it all the more bitter was that it was Angelique who had put her on to the scent. It was impossible to know—and probably just self-harm to speculate—precisely how long they had been tracking her, waiting for her to lead them straight to him. Could have been from the minute she walked out of their corporate HQ in the city; certainly long enough to know she'd been abroad, no doubt, from which they deduced that she had made contact.

Zal had begun packing his bag as soon as Holland's rentacops left the room.

'I'll find someplace else,' he said. 'Drop you a text once I'm settled. You better get to your meeting.'

Angelique would have shaken her head if it didn't feel like it was close to falling off. 'This isn't going to work, Zal,' she told him. 'You're here a day and they're already on to you.'

Her next line ought to have been that he should leave, that she couldn't ask him to do this any more, but she couldn't bring herself to say it in the face of an impossible choice. If he left, she would probably lose her parents. If he stayed, there was every chance she could lose all three of them.

Zal looked at her with an amused sense of minor grievance, like he was insulted by the idea.

'The last thing you need to worry about is those clowns,' he said. 'They shot their load too early. You only get one chance to take me by surprise, and they just blew theirs. There will not be a second.'

'They could get lucky, Zal. We don't know who else is out there. And given everything else I'm dealing with, I don't think I could take the irony of being the person who precipitates you finally getting nicked for the RSGN robbery.'

'Hey,' he told her softly, 'I always knew this was part of the package.'

He said it as though it was no more an inconvenience of coming here than the weather being cooler and the beer being warm. She remembered the look on his face when he was cuffed in the bathroom: rather than fear or panic, it was like: 'Sorry you have to see this'. But that was Zal all over the back, wasn't it? If he genuinely was afraid, the last person he was going to confide that to right now was Angelique.

At least she didn't have to torture herself with wondering what would have happened had she not gone back.

'Could you have undone those cuffs at any time?' she asked him.

'Not at any time,' he answered. 'Only when it was funny. Sorry, old joke. But there was no point doing it while they were watching over me with a gun. If you hadn't shown up, I'd have chosen my moment. In magic, timing is everything.'

She wanted to grab him, hold him. She wanted to tell him to come and stay at her place, her racing mind suddenly seeing a perverse logic to it being the last place they would now expect to find him: a triumph of emotion and desire over solid thinking. She just wanted to be near him, because everything felt better that way. Even if it was another of his illusions, she wanted to be lost in it, because it was far preferable to the reality.

She stares blankly, barely focused at the seated rows of reporters, the bank of cameras hemming them in at the rear. There's a clamour of questions, just white noise in her head, which is throbbing from tension. She places a delicate hand upon the swelling, traces a finger over the thin line of clotted blood.

They're dangerous for each other. That's why this can't work, why she was right to let him go five years back. With this thought, she realises that he hasn't asked her about that day at the musée, and promptly deduces why: he didn't go either. He knew it couldn't work. He had to run, had to find his new life.

Yet when she came and found him again, he

dropped everything for her. He loved her—for what else is love if not what he is doing for her now?—but she knows it can't work. Five years on, nothing has changed.

Her phone chimes, signalling a text and prompting the usual internal responses to which she seems incapable of developing an immunity. Too soon to be Zal, she guesses, and guesses right. It's another anonymous message. Her heart hammering, she presses the key to view it. It's just two short lines: one a statement, the second an instruction.

Having read them, she looks around at the cameras, the reporters, the cops, and nods to herself, for everything has become clear. She wasn't cut out for infidelity. Those two weeks with Zal five years ago had been the most exciting but also the least comfortable of her life, and right now she is being pulled to pieces by what she is being forced to do. She doesn't know how many people she is betraying, but she knows the figure is at least two too many.

She feels suddenly calm, the calm of resolution. She knows, finally, what she must do. No matter how it works out for her parents and the hostages, there are two other lives beyond that fate for whom she has a responsibility.

It has to end.

*   *   *

So what do they say about this state of consecration, this ascension to an exalted echelon of renown? What is fame? According to Byron, it is 'the advantage of being known by people of whom you yourself know nothing, and for whom you care as little'. Ooh, grumpy trousers, but spoken nonetheless with the admirable contempt of one who had already scooped his fill.

According to Socrates, 'fame is the perfume of heroic deeds'. What would the most venerated of Greek philosophers think of *Bedroom Popstars*, or *Big Brother*, were he to witness the willing indignities of those seeking only that perfume, and who regarded its scent as an end in itself? But perhaps old Byron was protesting too much, and perhaps there's nothing new under the sun. Tacitus said that 'love of fame is the last thing even learned men can bear to be parted from'. So perhaps one's deeds, whether heroic or in some other way remarkable, have never been entirely their own reward. And nor has that reward, that perfume, been dispensed throughout history with fairness or consistency, in any judiciously measured recognition of merit or dessert.

'Fame is like a shaved pig with a greased tail,' said Davy Crockett, 'and it is only after it has slipped through the hands of some thousands, that some fellow, by mere chance, holds on to it!' The honourable gentleman late of San Antonio, whimsical as his turn of phrase might be, is closest to the truth, particularly illuminating upon the arbitrary nature of this blessing as it is about to be bestowed upon my subject tonight.

I feel like Willie Wonka here, one last golden ticket to hand out. I hope its recipient is appreciative of how many people will be left disappointed once that last place is taken. Not merely because the lustre of endless, tantalising possibility always makes the announcement seem like a comedown, no matter how famous the celeb: an answer that can never be as exciting as the question. No, there will be disappointment also among those who have moved to exploit that excitement. The mere knowledge of there being one final celebrity to be abducted has itself provided a meal of carrion for the buzzards among the fame-hungry, those so ravenous as to have no concerns for their own dignity as they fall upon the maggoty meat. Such as those celebs who have, since yesterday, issued statements about going under protection—note ye, not merely gone under protection, but instructed their publicists to announce the fact—in order to stake a claim of being sufficiently famous as to be plausibly considered under my threat. And what is surely the greatest indictment is that more than one has had the same bottom-feeding idea.

Appalling, you may think, but there's worse: such as going into hiding, *not* telling anyone where you are and not answering your mobile, in order to get your picture on the TV news bulletins as 'feared missing'.

Oh yes. Step forward ubiquitous trash-icon 'Cassandra'. Real name: Sandra Clark. Occupation: having tits. The feminists slagged her off, but in truth, seldom has the male of the species been so effectively humiliated as by this self-seeking bitch's demonstration that a few pounds of strategically deployed silicon can blind them to an abject lack of charm, personality, intellect, talent or any glimmer of genuine sexuality.

And step forward disgraced former MP Liam Cadzow, along with, as always, his ghastly fucking harridan of a missus, Annabelle. Since he lost his seat in parliament and narrowly avoided the clink along with Aitken and Archer, I had thought that the only publicity avenue left unexplored by this pair was hardcore porn, but now it seems Liam has played a cute card by effectively saying, 'I'm Brian and so's my wife.'

I know exactly what they're all up to, being the one person who can say for sure that they haven't been abducted by me. They'll be keeping their heads down for as long as they dare, long enough to remind everybody of how famous and important they are, then they'll pitch up with some excuse about going off to the wilds seeking solitude. 'No idea all this fuss was going on. Slightly embarrassed, but it's very touching to know everybody cares . . .'

Part of me is curious to see who blinks first: Cassandra or the Cadzows; how many days they'd each be prepared to lie low, knowing the longer they're missing, the bigger the story, but that the first to break cover will scoop the most publicity. Unfortunately, I'm unavoidably about to pull the plug on the three of them by bagging my final public figure.

I'm standing in the hallway outside the apartment, my golf bag waiting on the floor, my invisible vehicle parked nearby. It's a pitiful little dwelling for an ostensibly successful professional at this stage in life, so telling to have an existence that can be comfortably accommodated within such a conveniently compact space.

It will be a simpler affair than the other three. I can't take any chances: this is not someone I want to be tangling with at close quarters, especially not a man in my condition. It's also someone more

421

likely to recognise me than Charlotte, Danny or Katie, so I can't allow my quarry the chance to scamper.

There will be no cameras or limos or dangling of bait. No publicists, chauffeurs, make-up artists or even hookers to worry about. Tragically, no lovers, partners or spouses either: the price of being married to the job, giving the best years of your life to your career so you're left eating M&S ready-meals for one in front of *Newsnight* before sloping off to your lonely bed. Still, look on the bright side. I will very soon be introducing you to some very interesting new people, a big chance for match-making.

That's half the appeal of these shows, isn't it? Who fancies who, who'd like to do what to whom, the vicarious excitement for the audience of seeing their own desires manifested by proxy, whether that might be sexual gratification or beating the show's most annoying cunt to death. All right, that's a bit more than previous celeb-reality productions have actually seen fit to broadcast, but if you ask me, that's where they've been going wrong.

It is pitch dark outside, a little after three in the morning. I consider it appropriate that I'm turning the police's favoured witching hour upon one of their own. They prefer to stage their snatch raids in the wee small hours when the suspects are sufficiently heedless of the impending danger as to be sound asleep, catching them unaware, disorientated and as unprepared to do a runner as anyone in their pants and stocking soles can imaginably be. I am appropriating their scheduling but modify my tactics as to eschew blue lights and

422

battering rams. A gentle finger on the doorbell will suffice.

*　　　*　　　*

Angelique is dreaming that she heard the doorbell and has got up to answer it. In her dream, she's in the hall, fully dressed but in her first police uniform, then in her primary school uniform, and the door she's approaching is now the one from her parents' first house in Leeside, in front of that swirly carpet James used to pretend was a pond spotted with lily pads. She hears the doorbell again, and this time the sound is enough to evaporate the dream and tell her that what prompted it was a genuine first ring.

*　　　*　　　*

It takes a couple of goes but I am patient. I know it was a long day for the poor dear, with much to keep a troubled mind awake despite the fatigue, too. I hear the sound of footsteps and conceal the gun behind my back, its tranquilliser payload prepped and ready.

*　　　*　　　*

Angelique lifts her head to look at the clock, feeling the swimmy-headed, extreme grogginess of being pulled from the most profound depths of a sleep it took a frustratingly long time to fall into.

The knackered part of her suggests she ignore it and go back under, woozily conjuring up images of drunks at the wrong door. Who would be looking

423

for her at this hour? This in its own fuzzy-logical roundabout way prompts the first coherent thought to strike her slowly waking brain: that it might be Zal. It's enough to get her sitting up, though she needs another moment to haul her body fully upright. The old t-shirt she's wearing barely covers her pants, but her intention at this stage is only to stick her head around the door sufficiently to establish the identity of her visitor. If it's him, she won't be worried about what she's wearing, and if it's anybody else, they can fuck off.

She opens the front door, still squinting as her eyes adjust to the brightness of the hall light after the dark of her bedroom. That it isn't Zal is the first thing to register. She finds herself looking into a male face. It is familiar, but in her bleary, half-dazed state and in this unexpected, dislocated context, it takes her a moment to recognise him. When she does, she feels a sudden, horrified fear, accompanied by the sensation of falling.

RESCUE

Angelique is not so much holding the door slightly ajar as holding on to it for balance. She can't bring herself even to speak, just stares helplessly at him, paralysed by this vertiginous feeling that her worst fears are about to be made reality by his word. The sight of Jock Shaw appearing outside her door at six o'clock in the morning would seem to precipitate only the most dreaded possibilities. A friendly face, a trusted mentor, a fellow Scot: of course he would be the one they sent. He's here to tell her that her parents are dead; or at best that the police now know they're missing and she is consequently off the case.

'I'm sorry about the early reveille,' he says, 'but believe me, mine was earlier. You better let me in.'

Angelique analyses his words and his manner like she's sifting through rubble for survivors. His tone is sombre, but the fact that he didn't go straight to it means it's not the death message. You don't preface that with a remark about yourself. Doesn't mean she's out of the woods regarding her secret being known, but anything north of the absolute worst feels like a relief. She feels her sense of self return, like she's been on the outside looking helplessly down on the scene for the previous few seconds.

'What is it?' she manages, squinting and accentuating the grogginess in her voice to disguise her reluctance to fully open the door. If she's busted, he's not going to say it to her out here, with her standing around in her underwear.

'The last mystery guest just signed in,' Jock tells her. 'It's Dale.'

Angelique gapes, then remembers herself enough to open the door. She stands aside to let Shaw pass, then uses the excuse of getting dressed to buy a moment alone in the bedroom. She switches on the light and glances at the crumpled duvet as she pulls on a pair of trousers left draped over the back of a chair. A few minutes ago she was asleep right there, and in the time since, she's been dragged abruptly to consciousness and forced to contemplate absolute loss before being rattled by a new sucker punch. Shaw's revelation didn't even allow her a moment of relief from the fears it had dispersed before instantly supplanting them with a new horror, one now enhanced with a generous helping of guilt. Her own dilemma remained in its stasis, oh whoopee fucking doo. And now Darcourt had Dale.

'Keen woke me up an hour ago,' Shaw explains, launching Firefox on Angelique's laptop, which he has booted up without seeking her leave. 'Your man Meilis alerted him around the back of four to say Dale had just shown up in Darcourt's final wee windae. He gave me the caretaker-manager role, told me to hold things together until we work oot where the fuck we are.'

'And what am I, club captain? Is that why I'm getting briefed ahead of the troops? You could have phoned.'

'You're on my route to HQ. But to be honest, hen, once I learned that thon bampot had taken Dale, I wanted to see for myself that you were all right. You've crossed swords before and he's as vengeful as he is deceitful. Just because he showed

426

just one window left to fill didnae guarantee anything. I was afraid he might be selling us a dummy.'

She recalls Zal's words yesterday: *Everybody's focusing on what's going to fill that window. Which ensures his next move will be something none of you sees coming.*

'He already has.'

Angelique looks at her laptop, where the embedded media-player has finished buffering and now shows David Dale pacing a cramped room in a sleeveless vest and jogging bottoms. He clenches his fists by the side of his head and lets out a scream of fury, but neither she nor Shaw hears it. There's no sound feed from the holding rooms, only the common area.

'Aye,' Shaw agrees, the word voiced in a bitter exhalation. 'He's been pissing all over us the whole time. Now he's taken the man leading the hunt, the public face of the investigation, and made his the next head on his trophy wall. He's telling the whole country the police cannae touch him.'

'Darcourt loves to sing when he's winning,' she agrees, 'but I think there's more to this, sir. It's not what he's telling the country: it's what he's telling us.'

'Which is what?'

'Check.'

Shaw looks reflectively at the screen, watching Dale strain against his confinement: he looks angry, frustrated, apt to start tearing the place up in his rage, but somehow silently and invisibly restrained.

'I hear you, detective. Meaning our next move could be the one he's relying on to set him up for

427

Mate.'

'No pressure.'

Shaw lets out a tired sigh of a laugh.

'I'm gettin' too old to be dealin' with this kinna pish, let me tell you. I've worked some no-win cases in my time, but from what I can see, this one is a rabid dug loose in a slurry pit. Anybody that goes near it is gaunny end up covered in shite, and that's only *before* they get bitten on the arse.'

'Welcome aboard, captain,' she says wryly. 'And thanks, by the way.'

'For what?'

'For your concern. For coming by in person.'

'Aye, well,' he says, a little too archly for Angelique's comfort. 'Truth be told, I wasn't without reason for my concern. Been in the wars, I see?'

The question diverts her before she can descend into further paranoid speculation. She puts a hand to her forehead. It's tender to the touch, but until Shaw made mention, she'd forgotten all about it.

'It's nothing,' she tells him. 'You should see the other guy.'

Shaw looks her in the face.

'I *have* seen the other guy,' he says. 'Or rather, I've had a phone call from the other guy's boss.'

Now she feels the lurching, vertiginous sensation return with a vengeance, the drop she was merely looking over the edge of when she discovered Shaw outside her door. She stares back and says nothing, which would have been a solid enough admission even before they abolished the right to silence.

'As the lady once said, there are three of us in this marriage, detective.'

428

Angelique swallows, buying a moment to consider her response and making sure her voice doesn't break when she gives it.

'Don't we have enough else to be worrying about?' she suggests heatedly, a dual gambit of deflecting attention back to other matters, while simultaneously implying that what he is referring to is not directly related *to* those other matters.

'Fortunately for you, as of a couple of hours ago, yes, I do have enough else to be worrying about. And if these were two entirely separate spheres, I would regard one as having massively superseded the other. However, this wee nippy sweety Holland at the RSGN said you came to her, sniffing for leads on Zal Innez, only a couple of weeks back. Yesterday he turns up in a London hotel room, in your company. Now, I don't know what was happening behind the scenes in Glasgow five years ago, and I made a point of not asking. But the shit we're up to our necks in now means that, if I'm making you club captain, as you put it, I need to know you're being straight with me. What's the Hampden, Angelique?'

She looks at the floor a moment, suddenly a wee schoolgirl in front of the heidie, a stumbling amateur who knows the most accomplished liars couldn't deceive a guy like Shaw. However, chucking him just a little truth might be enough to keep him off the scent of the real stash.

'If by "behind the scenes" you mean was I sleeping with him, then I'll admit it: yes I was. It might be early but I can't imagine that's the biggest shock you've had today. But that was, as you say, five years ago, and I hadn't seen or heard from him since. I tracked him down again because we were fumbling in the dark here and I thought he might know where the light switch was.'

'What, an armed robber?'

'No, somebody who can make *several* armed robbers disappear from a building surrounded by a hundred cops. Somebody who is an expert in precisely the kind of deceit, misdirection and outright showmanship that has so far allowed Darcourt to rip the pish out of us.'

'Well, could you not have asked Derren Brown instead of fraternising with a guy who could bring a pile of shit down upon us if this Holland woman gets what she's after?'

'He's my informant, sir. That's all.'

'Your informant? We're talking about the only listed suspect in an unsolved major armed robbery file—with my name at the top.'

'And would you have been happy to disclose the nature of your relationships with all *your* snouts down the years, sir?'

'Well, I never shagged any of them, if . . .'

Shaw cuts himself off with a sigh of frustration.

'We turned a blind eye back in Glasgow because Innez was the bait for a bigger fish. This time he's an unnecessary risk—to everyone, including himself. I'd advise you to advise him to do another disappearing act, because if I get the chance to collar him, I'm taking it. I'm a bawhair away from retirement and it might make up for being caught in the middle of this farce if I can sign off by closing another outstanding case.'

'With respect, sir, you never turned a blind eye because you never had eyes on him, never mind handcuffs. This won't be any different. He was smarter than us, he's smarter than Holland and I'd bet on him being smarter than Darcourt.'

'It's what you're staking on that bet that's bothering me, Angelique.'

'Needs must when the devil drives. But once we've got Darcourt, Innez will vanish again, for good. You won't see him before that, and afterwards, you won't need to worry about me or anyone else being compromised by the fall-out.'

'You sure about that?'

Angelique remembers her resolution of the night before. Coming from Jock Shaw, she couldn't ask for a sterner test of its fastness. There's a pang of regret, but it holds.

'I guarantee it.'

*　　　*　　　*

Oh, David, David, David. Chin up, old bean. Can't you manage a smile for all the lovely people watching at home? Dear, dear. Detective Superintendent Dale looks so down in the mouth, so humiliated. It's almost as if he can feel the wave of national disappointment that has greeted the announcement of his name as the final player in my new game. From those assuming it would be Cassandra or the Cadzows; to those maybe holding out hopes for a marquee name to top the others, such as a British-based Hollywood star or even a royal; to the lateral thinkers putting their smart money on a member of the cabinet: they're all letting out a collective 'Aw, that's rubbish', with only the third constituency perhaps nodding sagely to themselves in congratulation that they were on the right track.

I almost wish I could go in there and put an arm

around him, tell him not to take it to heart. I'm grooming him to be the star of the show, my most dynamic performer. These people out there have no vision, I'd tell him, and short memories too. It wasn't so long ago that they were making the same complaint of 'Who?' at the mischievous inclusion of non-celebrity Chantelle in the *Big Brother* house, only to vote her the winner a few weeks later. They don't know what makes a celebrity: they just accept it once they're told that you are one.

I'm watching Dale in the common area, where he has recently been introduced to Katie, Charlotte and DJ, and though he's already demonstrating his dynamism, he's not proving a good influence on his new friends at all. He's been thumping the walls to test their integrity and is now climbing on the furniture, a plastic fork in his hand, looking rather intent upon reaching one of the cameras I've got suspended from the ceiling. I anticipated this response to his being granted common-room privileges. The only reason I didn't warn him off this conduct in advance is that I was already planning to demonstrate to all of my guests what was expected of them by their host at this juncture anyway.

The widescreen TV on one of the walls is showing a countdown, with less than an hour to go. Perhaps that's what's got him so jumpy. The same clock appears on the website, as well as an inset on the several TV channels currently carrying the feed. They've got it on varying delays, Sky sailing closest to the wind with only ninety seconds, the Beeb stolidly protecting the nation's sensibilities as ever with a buffer of a whole five minutes.

432

I'm guessing whoever's got control of the cut-off buttons will be pressing them very soon. I hope they enjoy the accompanying buzz of justly executed authority, because I won't be allowing them this opt-out for much longer.

The clock on the big telly shrinks to a corner as the screen is filled with a feed from Anika's room. It shows the new patron saint of reality TV secured to the steel frame of her bed by handcuffs at each of her wrists and ankles.

'Detective Superintendent Dale,' I say, my voice booming around the common room from speakers embedded in the ceiling. 'I'm sure you'd rather not force me to rape the young woman you so movingly swore you'd do everything to protect. Please, play nice.'

And my, do they all sit quietly after that. Nobody says a word for the ten minutes or so that I leave the bound Anika visible to them, and there's precious little more gets said once the screen reverts to the countdown. The four of them are good as gold, sitting intently but impatiently round the telly, like kiddies waiting for *Watch With Mother*, anxious to discover what today's programme will be. Calibrating the scale, with *Camberwick Green* representing a short video announcing their imminent release, I'd estimate that what does follow is probably equivalent to the widely dreaded *Ring-A-Ding* and *Teddy Edward* double-bill.

The countdown reaches zero at noon.

I divert the video to all the feeds, so that it's the only show in town. It being a pre-edited package, the broadcasters won't stream any of it without seeing all the content, after which it's going to be a

433

tough call for organisations supposedly remitted to carry the news. Everybody is going to see it anyway, one way or another, and their nanny-state fixations with protecting their audience from offensive material are about to become moot to an extent they could not possibly imagine.

It starts with music, a drum beat fading up into my Nick Foster tribute mix: 'Hurts Like Dynamite', playing over a montage of images from the media coverage of Four Play and Vogue 2.2's abductions and subsequent retirements from the music business. This gives way to the SHE'S ALIVE! headlines that clamoured around Sally Smith's safe return, before Sally herself provides the segue to the ongoing plight of Anika with a recent TV message from her hospital bed, while 'Gone But Not Forgotten' fades up to replace 'Hurts Like Dynamite' on the soundtrack.

'You hang in, I know you're strong.'

I use this as the overture to a cloying compilation of headlines and TV clips about what a wonderful and special person Anika is, a recurring phrase punctuating the sequence with a growing frequency until it reaches a fast-edit crescendo: face after face, voice after voice saying the same thing: 'We'd do anything to get her back.'

The music stops and the image dissolves to show Anika chained to the bed, weeping. This fades to black for a beat, then fades up to footage of Dale at a press conference, stating solemnly: 'I will not rest, I will be doing everything in my power to ensure her safe return.'

Fade to black again, for a poignantly longer beat, then cue a slightly extended edit of my star-studded montage from the back of the limo.

'It could have been any of us, Anika. So we're not just thinking of you, we're *with* you. We're part of this. And we'll do anything it takes to get you back alive.'

'Anything it takes.'

'Anything it takes.'

'Anything it takes.'

'Everything in my power.'

Now, here's the part where the folks watching at home get to see something that the contestants don't. In the common room, the video ends there. For everybody else, there's an uptempo bass-drum beat, playing over constantly changing shots of all four faces—publicity stills, press photos and image captures from their respective feeds—one in each corner of the screen, with Anika's image fixed in the centre. The drumbeat is supplemented by guitars and vocals, my own remixed version of an old Echo and the Bunnymen number: 'Rescue'. There's umpteen songs I could have used with the appropriate word in the title, but I always hated those cunts and I know that its association with this is going to taint the track forever. No nostalgia-bandwagon greatest-hits release for you, McCulloch.

'Coming soon,' my own voice announces; no need for digital manipulation at this late stage of the game. 'From *all* responsible broadcasters . . . It's *Celebrity Rescue*! The show where four special people get the chance to prove they're as good as their word, and *you* get to see just how far they're prepared to go in order to save the nation's latest paragon: the lovely, the talented, the saintly, and the ever so vulnerable . . . Anika!

'It's the show where you get to find out just what

435

"anything" really means, and you don't need to worry about the action getting too hot for TV, because they'll be playing this game too, as will the police, the government, you at home; hell, we'll *all* be playing *Celebrity Rescue*, and here's why. Let me explain the rules.

'Every day, for four days, I'll be setting our rescue team a task, and if they complete them all, Anika will be set free, alive and unharmed. As will, in fact, all surviving celebrities, which if everybody behaves themselves, should mean three out of our four contestants. Why only three? Well, we'll come to that, and I promise, you're gonna love it! You're gonna love the whole line-up, because this is what you've really been wanting to see while you sat through those endless hours of brain-melting tedium on every other reality show!'

Three of the images fade out on screen, leaving only two flattering shots of Charlotte Westwood and Katie Lorimer.

'Task one will see these two sensuous beauties engaged in the hottest two-girl action ever seen on British television, and I don't mean the Wimbledon ladies' singles here. We're talking scissor sisters, we're talking oral, we're talking strap-ons and double-ended jumbo vibrators, ladies and gentlemen.

'But we're not going to let the girls have all the fun,' I add, as the pictures of Charlotte and Katie are replaced by images of Dale and DJ.

'With task two, we'll discover whether Danny "I'm all right" Jackson still sees the funny side of all those jokes about gay piano players pushing in their stools. That's right: it's celebrity buggery! With Detective Dale giving the naughty comedian a stiff lesson from his truncheon! But if DJ's feeling sore about being on the receiving end and not even getting the courtesy of a reacharound, then he'll be able to take it up with Detective Dale in a *real* celebrity deathmatch.

'That's right: task three will see our two male contestants armed with machetes in order to engage in a genuine fight to the death. Or you could call it a fight for life, because if either of them refuses to kill the other, then I'll be killing them both. But it would be no fun if that was the only incentive, which is why the winner gets the special prize of fucking both Charlotte *and* Katie— and we are talking ass-to-mouth here—in task four, our super celebrity threesome finale!'

The video now cuts to a montage of the contestants together in the common room, demonstrating the multiplicity of angles and close-ups I will have at my fingertips.

'And don't worry if you don't have access to the internet, because every evening, I will be supplying a one-hour highlights programme which will be broadcast on the BBC, ITV and Sky networks— after the watershed, of course, because Black Spirit Productions are a family-friendly company. If any of the networks fails to broadcast the highlights programme each night, unedited, then one of the contestants will die. And as with the massively popular *Dying to be Famous*, a similar penalty will be incurred by any attempted meddling from those nasty spoilsports in the police. If everybody plays by the rules, we're all going to have a lot of fun, and together, *everybody* is going to play their part in rescuing Anika.'

'And they say there's nothing good on the telly these days,' growls Shaw, breaking the dread-struck silence that has enveloped the Operations Centre. His words precede the inevitable cacophony of telephones, though Angelique can't see the point in anyone answering them. The purpose of any conversation is an exchange of information, and she doesn't believe two parties informing one another that 'No, we don't know squat either' really qualifies.

Shaw seems of similar mind, telling several receiver-gripping applicants to fuck off, or at least to convey to whoever is calling that he is presently indisposed.

'This part is like the electromagnetic pulse following a nuclear blast,' he says to Angelique. 'An overwhelming wave of energy serving no purpose but to bollocks up your communications network at the time you most need it clear.' He turns, waves an arm in a wide arc to get everyone's attention, then addresses the room.

'Our phones are for incoming information only at this time,' he announces. 'We're not answering anybody else's questions, just pursuing queries of our own.'

The phone next to Angelique rings at this insensitive moment, and she answers it principally with the intention of silencing the tone and quickly getting rid of the caller. However, it's Lee Hardacre from Met Traffic, getting back to her. The officers despatched to Dale's flat took a statement from a witness who heard a noise in the common stair and had a look out at the street to

438

see whoever was exiting. He said he saw a man dressed in black trousers and a black hooded top loading a long canvas bag into a dark-coloured van; looked like one of those holdalls for taking skis or golf clubs on holiday. He only saw the man from the rear and, predictably, he couldn't be sure about the exact colour of the van, the sodium streetlights washing all dark shades to black. What he could say for certain, however, was that it was two-fifteen, because he had glanced at his bedside clock as he returned to bed. That had given them a time-frame to analyse, and Angelique had relayed the details to Hardacre.

'We've clocked your van on two cameras in Crouch Hill,' he reports. 'Both sightings within a minute of each other, travelling west from DS Dale's address. It's a Ford Escort van, late-Nineties model, probably dark blue—'

'And you've checked all the cameras for any other possibles?' she interrupts. 'I'm not trying to tell you how to do your—'

'I understand,' he interrupts back. 'But believe me, this is the van.' There is something weary, almost apologetic about his tone, which Angelique doesn't find encouraging. 'The second shot was good enough to get a reg. I fed it through the system. DVLA records showed the plate decommissioned, but previously registered to a certain Mrs Mary Whitehouse.'

'Christ,' she mutters. 'That's Darcourt all right. I don't know where he gets this shit, but he's previously used fake plates belonging to Princess Di and Fred West.'

'He's a riot, isn't he?'

'So how long before the system picks up what

439

other cameras the van has appeared on?'

'That's just it. The results are all in. The van shows up on these two cameras then disappears. Which means either he's running this whole show from a location in Crouch End, or he switched vehicles.'

'Admittedly my London socio-geography is a bit rudimentary, but I can't see Crouch End as a viable spot for smuggling drugged celebrities in and out of a building unnoticed.'

'He'd need some budget for the square footage too. We're working on the switch theory. Problem is, there's no way of knowing when he started moving again, so the cameras are essentially useless.'

'What about the ditched van?'

'Got officers on the ground searching the area right now. I'll call you as soon as we find anything.'

While she was on the phone, Shaw has taken a call also, and looks less than pleased about it. Angelique deduces that it can only be from one person.

'Keen,' he confirms with a grumble. 'He's had the fuckin' home secretary on the line, asking the usual stupid questions politicians ask at this point. "Why don't you try such-and-such?" It never occurring to them that either we have tried it and it didnae help, or that there's a bloody good reason why not. In this case, it was why can't we trace the source. A sore point with the commander, I gather.'

Angelique recalls his face-off with Dale, Keen eventually backing down against his own instincts.

Meilis has latterly become sceptical about whether Darcourt truly has the capability to detect

440

attempted traces, and is increasingly of the opinion that the bastard actually had no prior warning before the police walked into his decoy honey-trap. However, he has continued to emphasise that the GOG technology meant that further honey-traps were the only thing they were likely to detect, and nobody was going to call that bluff twice.

'I take it Keen explained to the home sec that the reason we aren't tracing the signal is called Wilson Gartside.'

'No, he explained that the reason is called Sally Smith.'

Angelique nods to convey her understanding. Keen gave in to Dale on the grounds that Darcourt was going to kill all his hostages anyway.

'It wasn't the one he murdered that really fucked us,' continues Shaw. 'It was the one he let go. He played by his own rules, so we have to play by them too. Basic stick-up scenario: just do as he says and he won't kill anybody.'

'It's what he'll have them doing to each other that's gonna make it kind of hard to just stand back and watch.'

'Aye. Though not everybody will be so squeamish. The home sec will be speaking to the networks very soon, and there will be a lot of hand-wringing, heavy hearts and talk about having no choice if lives are at risk, but secretly the bastards will be thinking it's Christmas. And I mean Christmas as in the 1976 *Morecambe & Wise Christmas Special*: an all-time-high ratings bonanza.'

'Can't argue with you on that. The Great British Public will be vocal in their disgust, but they'll tune in in their millions. It'll overtake *The Teletubbies* as

441

the programme everybody sees but nobody admits to watching. God knows what the networks would be prepared to pay—in the public interest, of course—if Darcourt was auctioning exclusive rights rather than doling it out to all of them for free.'

'That's what I don't get, though,' Shaw adds, his voice dropping like it's a private thought expressed aloud. 'What *is* Darcourt's back-end on this? I mean, after everything else he's done, is his dying wish merely to put celebrity porn and gladiatorial snuff movies on the goggle box? Is it some kind of fucked-up moral revenge on the country that didn't understand him? What?'

'I know what you mean, sir. Appalling as it is, I must admit, it strikes me as insufficiently apocalyptic for it to be Darcourt's swan song.' She daren't mention it, but it's Zal's words that are resounding in her head once again, as she asks herself what they might not see coming while Darcourt's got them focused on this.

'And why didn't he announce when it was starting?' Shaw asks. 'He just said "Coming soon". He's got everybody where he wants them: what's he waiting for?'

The unlikely figure of Julian Meilis heralds the answer.

'Sir,' he says, standing up and removing his headset, itself a harbinger of great moment as Angelique can't remember ever seeing him *sans*; the thing is like a Borg implant. 'You have to take this call.'

'Who is it? Keen again?'

'No sir. It's *him*.'

With a gesture, Shaw silences the whole room.

'Darcourt? On the phone? Are you sure it's him?'

'It's not a phone call, sir. It's a P2P VoIP using AES encryption, routed directly to this desktop. Darcourt is the only person outside this room who knows the IP.'

Shaw stares bemusedly at Meilis, then turns to Angelique as if looking for some kind of explanation as to who this guy is and why he's babbling in another language.

'Julian,' she snaps. 'English.'

'It's a computer-to-computer call, sir, and he's one of the very few people who would know which number to dial.'

'Can we get a trace on it?' Shaw asks.

'Not a chance. That's why he's using it.'

'Can you record it?'

'Already there, sir.'

'Good. And run it through the speakers.'

'You got it.'

Shaw takes the offered headset and holds it awkwardly, like it's a live crustacean, turning it until the mike is in front of his mouth.

'Hello?' he tries tentatively. 'Hello? This is Shaw. Can you hear me? Over.'

Angelique catches sight of Meilis rolling his eyes at this; he's very lucky Shaw didn't.

There's an expectant silence, lengthening to the extent that Shaw starts scrutinising the headset. Then Darcourt's voice booms all around the room.

'Detective Chief Superintendent Shaw. That's quite a mouthful for the troops, *n'est-ce pas?*'

Angelique finds herself putting her hand to her chest where that shotgun blast impacted. It's him all right. Clear as a bell, unsettlingly familiar, in that weird accent. There's very few traces of his

443

native Scotland left, all smoothed over within a voice that has been speaking another language for years. She knows the effect well: some of her own pronunciations have altered as a result of her time in France, but in Darcourt's case, it must be exacerbated by the effort of pretending to be someone else every minute of every day. Aware of her own tendency to switch between at least two, and sometimes more, she wonders what language he actually thinks in. She decides Prick.

'I find they manage fine with "sir",' Shaw responds. 'Listen, I'm glad you called. You never said what time your programme was on at, and I was wanting to set the video. Case it clashes with *Corrie*, you know?'

'Mmm,' Darcourt mumbles dispassionately. 'It would be a tragedy if you missed it, I agree. And I hope that's a figure of speech and you're not really making do with some steam-powered old VHS. A television event of this importance warrants Sky-Plus at least, though a proper hard-disc recorder would be . . . decorous.'

'Aye, I might splash out. I don't suppose you've plans to release a DVD box set? Nah, you've been generous enough already, laying this on for free.'

'I call it public service broadcasting, in the spirit of the late John Reith. This sexually repressed little backwater of a country needs to undo a few buttons and confront reality. The British zeitgeist is that of an entire people that has collectively never got over walking in on its parents having sex. I've heard Katie Lorimer called the unofficial mother of the nation, so I'm guessing if the nation sees Mummy playing an away fixture with the England captain's wife, then its psyche will be

444

forever altered for the better. Wouldn't you agree?'

'Not entirely. I'm a bit concerned about getting woken up in the small hours by the same wee nation because it cannae sleep for having nightmares since seeing its Saturday teatime favourite being raped and then hacked to death.'

'So your money's on Dale for the big fight? Yeah, mine too. Banking on it, really. Don't imagine there's many people want to see Danny Jackson humping either the WAG or the MILF. They want a fit, strong specimen like the good police officer. That's the beauty of it, in fact. If everything goes to plan, the only person to die will be nasty, pudgy, racist, queer-bashing old DJ, and everybody probably figures he's the most horrible and therefore kind of deserves it. They'll make their peace with it, because the prejudices inherent in their supposed liberalism are really just another form of bigotry. Like I said, it's a public service: the public need to hate guys like Danny Jackson to make them feel better, same as they need to hate me. Even more generously, I will be conveniently mopping up all their guilt for them, because I'm the one going to hell: they're only watching.'

'Aye, very good. Your OBE's in the post. Now how about you quit twanging your wire and tell me what it is you called to ask for, because I'm getting a strong niff of "want" blowing in from your direction.'

Darcourt lets out a derisory little laugh, like this is all so beneath him.

'Yes, very erudite, officer, and indeed very perceptive. But it's not about what I want. It's what I can give.'

445

'You can give me the five people you're holding hostage. That's all I'm interested in.'

'Oh, but don't you see, I'm holding hostage something of infinitely greater worth than the lives of five people, no matter how much more valued they seem to be above the average mortal. I'm holding hostage the nation's sense of its own moral security. What do you think it's worth for the nation *not* to see its Saturday teatime favourite being raped and then hacked to death? What do you think it's worth for the people of Britain *not* to see Charlotte Westwood tonguing Katie Lorimer's clit before doing her doggy-style with a dildo? Is it possible to put a price on these things?'

'I've an irrational hunch you're about to. How much do you want?'

'A hundred million pounds.'

He says it instantly: no pause, no build-up, no hesitation. His tone asserts that it's not merely a high gambit to open a negotiation.

'Aye, tell you what, just give us two minutes here and the boys at the station'll have a whip-round.'

'I don't believe that will be necessary. I think Her Majesty's Exchequer will be prepared to cover this one. In fact, I'd be very surprised if senior government figures haven't already been in touch, and as this should remain our little secret, they need have no qualms about negotiating with hostage-takers.'

'They might have one or two qualms about doling out a hundred million sheets.'

'Hardly. It's a figure their spokesmen are happy to admit to losing all the time. "This week's two-day postal strike is estimated to have cost a hundred million pounds". "The transport disruption caused by Tuesday's unexpected snowfall has cost the economy a hundred million pounds". "An outbreak of diarrhoea at several firms in the Square Mile is reckoned to have cost . . ." guess what. It's always a hundred fucking million. It is also, poetically enough, roughly what the government has taken in taxes accruing from reality TV since the first *Big Brother* was broadcast in the year 2000. Ill-gotten gains, about to be confiscated. Otherwise they get my reality TV show, gratis.'

'All this build-up and it turns out to be nothing more than a shake-down? I have to confess I'm a bit disappointed.'

'Only because I haven't shown you the full, deluxe, national peace-of-mind package. I'm not only offering you the hostages. I'm offering *myself*, Detective Chief Superintendent. You get to catch me. Justice done, threat eliminated. British bobbies get their man, so we can all sleep safely again, celebs and mere mortals alike. You'll also get my cooperation, once I'm in custody. I *will* be telling tales, and believe me, I am a fount of information regarding certain individuals and organisations. What I can tell you is worth a hundred mill on its own. Christ, this whole deal is a fucking bargain now that I'm adding it up. But then I suppose it's kind of a closing-down sale, because everything must go: including, sadly, me.'

Just in case Angelique isn't quite feeling tense enough right now, her mobile chimes with the arrival of a text. She daren't even look at the phone to verify who it's from while there's anyone close by who might be watching. One glimpse by another cop of a new proof-of-life picture and she'd be ruined. She's desperate to look, though, hoping for the temporary peace of mind it would bring were the text *not* to be from them. She felt the lurching sensation the moment Darcourt said he was prepared to tell tales, and when the text alert sounded, it seemed such an omniscient reminder that the paranoid part of her wondered just how close these 'unknowing' sources might be.

'Your silence is touching,' Darcourt resumes. 'But there's no need to feign surprise. I've seen that photo of me you released a few days ago, and I know where you must have got it. So if you've been accessing my private medical records, then you've scooped me on my big revelation, haven't you?'

'We're all really gutted for you,' Shaw says, confirming.

'Oh, don't weep for me, but for yourselves if you don't take what's on the table. Because otherwise I will deliver my promise to the viewing public, then fade away, and there will be no resolution, no closure. I will be taking myself off somewhere to die, quietly and comfortably, never to be found, a death never to be certified, which means you will never be free of me. I'll always be haunting the country, haunting you cops as I cloud the public's imagination. After every murder, every abduction, the question will be posed: was it the Black Spirit? Is he really dead? And my parting broadcast will inform the public that I offered this deal but the authorities said no. Now, isn't that hundred million starting to sound like a smaller and smaller sum?'

'Sounds like a steal, but then it wouldnae be my money. And I really have to ask, what do you need a hundred million for when you don't have a lot of time left to spend it?'

'That I don't have long left is why I need it. There are some things I want to ensure for the future, even if I won't be around to see it.'

'You mean your son. Connor McRae.'

'Things will be difficult for him. There's no permanent way of shielding him from the truth, not with our media, but all of life's burdens are easier to carry when money is not an issue. I've taken precautions. He won't be able to identify where the money has come from any more than you will be able to trace where it went.'

'He'll be able to work it out eventually.'

'And then he can make his own choices. But having done nothing wrong himself, he may not find it morally imperative to give it back.'

'A hundred million is an awful lot for one wee boy. You not worried about spoiling him? I mean, look what a prick his father turned out.'

'Indeed, and the sins of the father must not be visited upon the son. The money won't all be going on him, however. I want to leave a legacy: set in motion a few things that might cushion the blow once he finds out the terrible truth about dear old dad.'

'I reckon you'd better set aside ten million just for his therapy.'

'You'd be surprised what worthy causes I'm setting money aside for, Detective Chief Superintendent, but it's all moot unless I get it.'

Shaw sighs, looking to the ceiling, perhaps because it's the only place he won't see either a monitor or a colleague's expression. He looks like he wants to punch something, the body language reminding Angelique of Dale's inside that common room.

449

'So how would this work?' Shaw asks.

'It will work according to my instructions: that is the first thing to implicitly understand. There will be no simple handover or exchange, unfortunately, but this is for your—or rather the government's—benefit. Nobody is going to believe that I just suddenly turned myself in and freed my hostages, and the authorities can't be seen to have made any deals with a monster like me. To that end, you brave boys in blue are going to catch me in the act.'

'I'm not sure how plausible that's going to look either, given how we've managed so far, but I'm all ears.'

'It will be all too plausible. Two nights from now, you will stage a concert at the Tivoli nightclub. I assume I don't need to underline the significance of the venue. The event will be billed as a public gesture of defiance, a tribute to my victims and a show of solidarity with my new hostages. It will feature a star-studded bill. I'll leave the line-up to you, bearing in mind that it will only conceal the government's collusion if you get a decent quorate of celebs. A room full of coppers won't really sell it.

'Thus goaded and tempted by so many famous faces, in my greed and insanity, I can't resist trying one more daring raid to scoop the pot. But alas, it turns out to have been a trap, and you slap the cuffs on me. Gorblimey, it's a fair cop guvnor, I'll come quietly. In front of the cameras too. What a result.'

'What about the hostages?'

'Once I'm in custody, you will disburse the funds and allow me to verify that the transfer has been made, using a laptop. Once verified, I will take you

450

to the hostages. However, before I leave for this little soirée, I will be setting a timer to deliver into each of the hostages' holding cells a massive dose of the gas that killed young Wilson. The countdown will be only a matter of hours. You should therefore not attempt to mess me about with any "unforeseen" delays, and you'd better make sure the traffic is clear too. Any questions?'

Shaw thinks for a moment, his mind already calculating logistics.

'Yes,' he declares. 'Once you've verified the transfer, what guarantees us that you will honour your end? You can just sit there with your thumb up your arse. Christ, you can ask for another hundred million. The clock will be ticking, after all.'

'I have to confess, I have often wondered why anyone would want to, and indeed whether anyone ever *has*, sat around with their thumb up their arse. No matter, allow me to clarify: at this point, the funds will be in my account, but eminently recoverable. Once I have taken you to the hostages, however, I will be permitted further access to my laptop in order to execute a number of electronic transactions, and that way we'll both have what we want.'

'And once you've taken us to the hostages, what guarantees you that we will honour our end?'

'I trust you,' Darcourt replies. 'You're the police.'

'Imagine for a moment that I'm one of those bad apples you hear about.'

'Then that would be all the more reason not to answer your last question. Now, are we clear?'

'Not quite. It's very short notice. Two nights may

451

not be en—'

'May not be enough time for you to explore your options and dream up some ingenious scheme to get hold of me and the hostages without paying. It's a party in a nightclub, for fuck's sake. The only thing easier for you to organise would involve a brewery. It takes place two nights from now, and if it is suddenly postponed for any reason, then I kill a hostage and we schedule it all again for the next night. You got that?'

'Yes,' Shaw says begrudgingly.

'Good. Now go and talk to whoever you have to talk to. Get me my money.'

'How do we get back in touch?'

'You don't have to. I'll get back in touch with you. But in the meantime, the way it works is this: if I don't see mention of the great "Get It Up You Darcourt" party on the news by lunchtime tomorrow, then I promise you, celebrity muff-munching will be the lead item on the evening bulletin.'

THE INVISIBLE PASS

'They'll go for it,' Shaw says to her, once Keen has departed for the emergency meeting with a copy of Darcourt's call on disc and Shaw's analysis to back up his briefing.

'What if it's a double-cross, another trap?' she asks. 'Get even more stars—not to mention a whole bunch of cops—all in one place, then wipe out the whole lot with a bomb?'

'We'd have absolute control of the nightclub, for forty-eight hours in advance.'

'Darcourt had access to the same premises several weeks back,' she reminds him. 'And he tends to plan ahead further than forty-eight hours.'

'There wouldn't be anything in that for Darcourt. If he wanted to take out more celebs, he could do it like he's done all the others. There'd be no need to offer this deal.'

'I think you're right,' she agrees, thinking of the Baker kidnap, the Lombardy incident, and of Darcourt's living legacy. 'His time's running out and the money is the only thing that would extend his reach beyond the grave.'

Sitting shortly afterwards in a toilet cubicle where no one else can see her, Angelique's fingers are shaking as she finally gets the chance to type her response to the text that arrived during Shaw's call with Darcourt. She's spent so long in a suspended state of apprehension about what might happen if it came to this, all the while just as afraid that it would never get so far. Now it's real, and she's setting it in motion the second she presses

453

Send.

The alert was from who she feared, albeit the content at least failed to support her more paranoid thesis. They had seen Darcourt's 'trailer'—who hadn't?—and with him having made his next move, they wanted an update.

What an update.

This isn't just betrayal, it might well be treason. Senior cabinet members are still discussing Darcourt's deal a couple of miles away in Whitehall, and she's not only feeding criminals the same details, but informing them of the decision she knows the government will make before the ministers themselves have made it.

Christ. She remembers the first time she saw this city, on a midterm long-weekend trip with her parents in 1986, getting Mum and Dad to herself for a change, with James opting to stay in Glasgow now he was old enough to be left on his own. Mum insisting they visit Harrods. Dad predicting, accurately, that all she could afford to buy was a souvenir plastic bag. Downing Street. Westminster. The thrill of seeing these places for real; the wonder of what could be going on right then within those walls.

'Aye,' Shaw had concluded. 'He wants that money, and if you ask me, he's going to get it. He's been smart enough to make it appear the path of least resistance to the politicos. He's done all he can to guarantee both ends. As long as they're confident we can pull off the exchange and not get stiffed, they'll give it the nod. Politicians care more about perception than reality. He's set it up so that everybody's going to come out of this looking like winners, but it will all be an illusion.'

454

Trembling in her locked bathroom stall, Angelique is fighting back tears, aware that once she presses Send, then her own illusionist will have to find a way of disappearing Darcourt without trading five other people's lives for an outside shot at saving just two.

*   *   *

Zal takes the call as he sits in a booth in a large, airy pub on Euston Road, back in the room so as to be secluded, but with a clear enough view through the windows as to be able to watch the station entrance. It's Angelique, speaking over the sound of traffic, calling from some place outdoors because she can't talk where she might be overheard. She sounds like she's barely holding it together as she fills him in on the concise facts of the deal, filling him in also on the fact that Jock Shaw now knows he's around. This troubles Zal the way an electric bill troubles a guy who owes the mob fifty large.

'A hundred million,' he echoes, mulling over Darcourt's audacity. 'My old man used to say it was no use being the richest man in the cemetery, but I guess he was wrong. Look, I need to work on this. We'll talk about it tonight, someplace safe, when you can tell me more.'

'Tonight? Where?'

'I'm working on that too.'

*   *   *

Zal sees them only a few minutes later. They've come early, intending to beat him to the station,

455

where they'll attempt to remain unseen while monitoring his approach. Too bad he's already done this to them.

He finishes his drink, wheels his suitcase across the dual carriageway and proceeds inside Euston Station. He doesn't know where they are, doesn't look for them either. Instead he progresses casually through the concourse, stopping in front of the departure board, where he pulls out his ticket and quite unnecessarily cross-refers the information. Even these two assholes can't have missed him now.

Having advertised himself thoroughly, he wheels his case towards the appointed platform and has his ticket checked by the guard. He makes a point of asking the guy which coach his sleeper berth is on. It tells him on the ticket, but the point is to draw attention to his name, or at least the one he's travelling under.

Zal drags his case almost the entire length of the platform to the corresponding coach, two from the front. He doesn't have rear-view mirrors attached to his head, but if he did, he's pretty sure that right now he'd be looking at Bullet-Head and Comb-Over roughly fifty yards behind him.

He called his old friend Jerome last night, once he was sure he'd given them the slip by taking a cab from outside the Halton Court Hotel. Zal paid the hack upfront to keep driving while he slipped discreetly away during a stop at a red traffic light, before disappearing into the tube network. He remembered Jerome had played in London a couple of years back with the experimental theatre group he helped found, using what he liked to refer to as 'a grant' from the RSGN. Zal needed an actor, somebody Jerome could not only source at short notice, but seriously vouch for too. The only other qualification was that he had to come across as plausibly heterosexual. Zal explained this as a belated but crucial afterthought, prompted by Jerome's familiarly hammy tones. Zal and a select number of women knew for certain that Jerome wasn't gay, but you'd have a hard time convincing a stranger of it.

He put Zal in touch with a guy called Maddox, endorsed by Jerome as being 'from Newcastle, which would mean it was impossible for him to sound faggy even if he was as queer as Leo'. It was reassuring that in an ever-changing world, there were certain things you could always rely on, such as Jerome getting a dig in at their mutual friend Leo, even if Leo wasn't around to be insulted by it.

Maddox was ideal: mid-thirties, in good shape, attractive but not pretty, and a convincing performer. Plus he was broke, which helped, though given his profession, this hardly constituted Zal catching a lucky break.

Zal gave him Bullet-Head's mobile number, which Angelique had noted before returning his phone. Following Zal's instructions, Maddox told the ex-cop that he knew where his quarry would be that evening, but it was going to cost him: a thousand in cash.

'Who the fuck are you?' Bullet-Head asked.

'I'm Angelique's boyfriend. Or at least I thought I was until this cunt showed up. Now I don't know quite where I stand, and she's in a right state. I want him out the picture.'

'Shouldn't you be paying me, then?'

'Listen, I'm a man who pays attention during

pillow-talk. I know about this guy's history and I know what he's worth to you, so if you want him, you fuckin' pay us, and then I can maybe get the lass something nice to help her get over her disappointment.'

'All right. You tell us where he's gonna be, and if it leads to—'

'Aye, very funny pal. I could get that offer from the polis. I'll tell you the gen when I've got the cash in my hand. I'll be in the Green Man on St Martin's Lane for one hour. After that, who knows?'

According to Maddox, Bullet-Head and Comb-Over turned up inside half that, bearing half the money. 'This is all we could get at such short notice,' Bullet-Head explained, figuring Maddox would take a bird in the hand.

As per Zal's instructions, he told them to fuck off.

'Better get to the cash machine then, and hurry, because this information is only good for another two hours.'

With an ultimatum added to the equation, Bullet-Head sent Comb-Over out to the street with all their bank cards and PINs. He returned in about ten minutes with another five hundred, which was when Maddox informed them that Zal was booked on to a sleeper train to Glasgow under the name Alex Harvey. 'Going up to Angelique's turf to lie low.'

'The dog returning to his vomit,' Comb-Over apparently opined.

Zal parks his suitcase inside the cramped sleeper berth and quickly performs his preparations. There are two bunks in each little compartment. Zal has paid for four, an even greater extravagance when he considers that he won't actually be going anywhere.

He exits with haste rather than hurry, then locks himself inside the second compartment he booked, two doors along. As always, it's all about the exit, and this time, he's out before they'll even arrive where they expect to find him in. And therein lies the factor that's screwing with his thinking on Darcourt: no matter whether you want to call it ransom, extortion or whatever, what this psycho is pulling is still, in the end, a heist. You plan a heist backwards, starting with your out—but the problem is, Darcourt doesn't need one, because he already has the biggest out imaginable, one nobody can stop him reaching. This allows him to bypass all the normal considerations. They can't threaten the guy because he doesn't have anything left to lose. The only thing in his world now worth caring about is the money, which the cops will be caring about a great deal less than five hostages, not to mention collaring the man himself as both a prize and an invaluable source. According to Angelique, their financial consultants are confident that the money can still be traced as long as they've got Darcourt in custody, with clawing it back made easier once the cancer finishes him off. They're aware the guy must have tricks up his sleeve, but they figure it's well worth the risk considering what's there to be gained. Money can be recovered, but any time Darcourt has been involved, they haven't recovered people: just bodies.

The plus side of this deal is that Zal's task is now simplified to spiriting Darcourt away from the cops; albeit half the cops in central London. Angelique can get him the nightclub layout and floorplans, as well as all the information he needs about the police deployment. That said, maybe the place Darcourt leads them to once he's been apprehended might prove better for springing him. Zal wouldn't have any advance notice of the location, but neither would the cops. Plus, as Angelique explained with some anguish, spring him too early and it leaves five people to die.

Unless, of course, there's some way of getting Darcourt to lead Angelique and Zal to the hostages before taking him to trade for her parents. Unfortunately, the problem with that would be the same one the cops can't figure: even if they make it to where the hostages are being held, Darcourt has an unseen card to play that guarantees they can't stiff him. Zal's guess is it must be a means of stopping the gas and releasing the prisoners, which somehow only he can execute. A code? A password? No. It's a black-bag job. He might be dying, but with five lives at stake and no public knowledge of this operation, the cops could torture him in any way they can imagine in order to get that information. Darcourt knows this, so he must have some other safeguard, but what?

Zal's musings are interrupted by the sound of Bullet-Head and Comb-Over knocking on 'Alex Harvey's' door, just down the corridor. They'll have asked the guard on the platform, or maybe the train's purser, which compartment Mr Harvey is berthed in, maybe even flashing fake or expired ID if necessary. They give it a couple of shots, one putting on a different accent and claiming to be the ticket inspector. With Mr Harvey proving extremely reluctant to answer the door, they resort to barging it. Zal hears a crash, which proves more immediately efficacious than its proponent probably expected, but then the asshole should maybe have tried just turning the handle first.

Once inside, Zal knows they will find his suitcase on one of the bunks and a brown envelope, addressed to himself, sitting on top of it. 'Where the fuck is he?' he hears Bullet-Head ask through the card-thin walls.

'Maybe he's in the fuckin' suitcase. I dunno. Probably gone to the bog.'

'Let's see who's been writing to him, eh? Oh, fuck . . .'

There ensues an outbreak of coughing and spluttering, as whoever opened the booby-trapped envelope released a cloud of chemical irritant that will have engulfed both men inside such a restricted space. Fortunately, there are two bottles of water sitting handily on the narrow shelf-cum-table built into the wall opposite the bunks. Less fortunately, both bottles are laced with fast-acting sedative.

Zal hears the thumps about thirty seconds later, and makes his way back along the corridor. He removes gags and plastic restraints from his suitcase, then sets about securing the now sleeping ex-cops to the bunks. It's a bit of a heft getting Comb-Over on to the top one, but Zal manages it. He gags them and ties them with their faces turned to the wall, then drapes the blankets over each. If the purser sticks his head around the door, he'll find them both sleeping peacefully, their tickets—to the train's final destination—lying out for inspection on the table by way of saying 'Do Not

Disturb'.

Zal closes the door gently and alights from the train. He stays on the platform until it pulls away, making sure nobody gets off. He hopes they enjoy Inverness.

He walks back towards the station entrance, allowing himself a moment's satisfaction at his night's work, with the thought of a far greater task looming ever closer on the horizon. Maddox sold it well, a truly convincing performance, though the selling part itself was what guaranteed its success. Not leaving until he had that full thousand had been Zal's most emphatic instruction. He didn't just need to get them to pay for the tip-off, he needed to get them to pay dearly for it. It was one of the most enduring truths of human deception: the harder it is to come by certain information, the more credibility you attribute it.

Reflecting upon this, he is suddenly struck by something that stops him dead on the platform.

'Ain't *that* the truth,' he says out loud.

*　　　*　　　*

Angelique catches a glimpse at her watch as she climbs the stairs to the serviced apartment, and it makes her head hurt just that bit more to see that it's after eleven. What time was it when Shaw woke her this morning? She's too tired to do the arithmetic and work out how long she's been on duty. Fatigue is good, she tells herself. If she's sufficiently exhausted, she thinks, maybe it'll help her sleep, but shit, who's she kidding? Nitrous oxide wouldn't help her sleep right now. It's like she's got so many things to worry about, she can barely decide which one to concentrate on, so there's just this chaotic, mobbing assault like her fears are birds and she's Tippi Hedren.

The one that alights for a brief peck as she acknowledges the hour is that Zal never got back to her. It was largely moot, as she hadn't got free until forty minutes ago, but that he didn't get in touch provokes fears that Holland's goons have got him, to say nothing of Shaw's own threat. Could be he has just been playing it safe and didn't want to call her on her mobile. Maybe he left a message on her machine.

She puts her key in the lock but it doesn't turn. She gives it another twist, feeling just too bloody tired to be doing with this. It's the one last crappy, insignificant thing that could nonetheless tip her over into pathetic, sobbing helplessness. However, this danger is rapidly dispersed amid the adrenaline sharpness that accompanies her noticing light through the keyhole when she pulls it clear of the lock ahead of a fresh try. She came out in a rush this morning, she reminds herself, could have left a light on in her hurry. She honestly can't remember, but either way, the sharpness gives her the clarity to realise why the key wouldn't turn: the door is already unlocked.

Now, she may have been in a rush, and she may even have left a light on, but she definitely wouldn't leave without locking up, especially not in this town.

She pushes open the door and takes a step back into a defensive stance. It's the hall light that's on, and looking along it she can see that the living room is lit too. A human shadow is visible on one wall, the silhouette distinct enough to answer all

463

her questions.

She closes the door and locks it, then proceeds to the living room. Zal is seated at the apartment's tiny dining table, a pack of cards before him, cut into two piles. The other chair is positioned directly opposite. He gestures to her to take a seat, his hands poised over the split deck.

'Zal. I won't ask how you got in, but I have to inquire, what the hell are you doing here?'

'I said we'd talk someplace safe. I figured this place was as safe as any.'

'Are you kidding? What about Holland's numpties? They know I live here.'

'Yeah, they could be a big threat, whenever they get back from their trip to the highlands.'

He grins, the same grin he gave her five years ago, hinting at mischief and schemes and secrets that it is in his gift to divulge or withhold. It could be her own desperation, could even just be that she's too damn tired to consider the glass half-empty, but she detects there's more to this smile than mere pride at having dispatched two more bumbling pursuers.

She sits down, facing him across the table. 'So let's talk,' she says.

Zal picks up both halves of the deck and riffle-shuffles the cards, then places the complete pack face-down on the table. He taps the back of the top card twice, then turns it over. It is the eight of diamonds. She smiles in patient acknowledgement, looking him in the eye. It's their card.

'There's a number of card tricks performed using a technique known among magicians as the Invisible Pass,' he says. 'It's a standard sleight that secretly cuts the deck, but you gotta do it amidst a

464

bit of distraction for the audience, because it's hard to conceal if their attention is focused directly on the pack. Conventional wisdom dictates that you can't—or at least you shouldn't—do it close up like this, what's known as table magic. Now, when my dad was learning the trade, in Glasgow variety theatres way back in the day, his greatest mentor was a guy who could do the Invisible Pass as close as we are now: one-on-one, just a table and a pack and no distractions. He taught my dad, and my dad taught me, but the lesson to be learned is not the technique.'

Zal cuts the deck and places the eight back in the pack, then taps the new top card, revealing it to be the two of clubs.

Zal tugs back his cuffs and flexes his fingers, bending his head over the table.

'I will now bring the eight of diamonds back to the top using the Invisible Pass. Do not take your eyes off the deck.'

Angelique focuses on the blue-backed stack, trying not to be distracted by albeit minor movements of Zal's fingers just either side. With a sudden deliberateness, he waves his right hand over the pack, briefly concealing it from view but not close enough to touch it. Then he lifts his head and smiles.

'Did you spot it?' he asks.

Angelique looks up at him and blurts out an involuntary laugh. The words 'No fucking way' leap to mind. Zal's eyes direct her gaze back to the table, where he lifts the top card, confirming it to now be the eight of diamonds.

'How the hell did you do that?'

'I told you, the lesson is not about the

technique.'

'So what is the lesson?'

'One you need to understand for yourself rather than have spelled out. But if I was to give you a hint, it would be that your question should not have been how did I do that, but *when* did I do that.'

She gets it.

'You did it when I laughed, thinking the trick was done.'

'You make your move on the offbeat, always.'

'Cute, Zal,' she concedes. 'But you'll have to elaborate a bit with regards to how this principle applies to the matter in hand. You know: the one about vanishing Simon Darcourt from underneath the gazes of about a hundred police officers, many of them armed with semi-automatics.'

Zal assumes an expression of butter-wouldn't-melt sincerity.

'Have you thought of just asking them *real* politely?'

Angelique resists the temptation to pick up the deck and throw it at him, an impulse made easier to restrain by the assumption that his taking the piss indicates he's happy about something.

'Just working on the improbable hypothesis that your ingenious first suggestion has an infinitesimally tiny flaw, would you by any chance have a Plan B?'

NONE OF HIS TRINKETS WANTING

And now, the end is neah . . .

Can't get that fucking song out of my head at the moment, and to make it worse, I mean the Sid version, boasting the quite unsurpassed crassness of rhyming those immortal opening words with the line: *You cunt, I'm not a queeah!* God almighty, make it stop. I can't let that be the internal soundtrack to my farewell performance.

Understandable that it should be something theatrical and hammy, I suppose. Here I am, after all, sitting in front of the mirror, putting my face on while I wait for the Final Act Beginners call. It's odd, though, that I should feel so nervous, as it's many long years since opening night for this old-stager. Maybe all the veteran thesps feel the butterflies returning when they know it's their last ever show. The pressure for it to be perfect is all the greater because there can be no making up for it if it goes wrong, and though it may not be the show they remember you for, you want to tell yourself you were still able to give your best, even at the end. I want to give it my best. I want it to be very special, as befits the task in hand: that of executing the one, ultra-high-profile death warrant that really did come in a manila folder.

That's why I've been double and triple checking everything for hours: timers, sensors, triggers, gas levels, engines, fuel, batteries, phones, computers. After so much planning, so much painstaking and meticulously executed work, I'm getting cumulatively jumpy about the possibility of a single oversight or act of forgetfulness making it all for nothing.

It's only nerves, though. I've been methodical and thorough. My systems are in order, my prisoners secure in their holding cells, all having succumbed to the sleep agent I have deployed in case anybody might get boisterous and deduce from my lack of a response that the cat's away.

There remains but one final duty. I have to record my parting-message video, as every good suicide bomber must. It's the only thing I haven't fully thought through, as even this close I remain unsure what to say. Penitence would be appropriate, much as it might stick in the throat, but I have to remind myself that it's about what I intend to leave behind, not how I feel right now. I have to give them some kind of closure. I owe them that much. I have to remember that every act of contrition, every gesture of amelioration, while gall to my lips, may prove balm to the son I will never know, but who will unavoidably one day know me.

My make-up complete, I take a last look in the mirror, the final time I will see this face. Then I turn my seat so that I am facing directly into the video camera, and I set it recording.

*   *   *

Standing in the musty half-light of his old lock-up, Albert turns the nightstick in his grasp, his fingers reacquainting themselves with its touch, like he's shaking the hand of an old friend he hasn't seen in years.

'Well, Mr Spank,' he says, 'can you also be

coaxed out of retirement for one last ride?'

Though he hasn't been inside it for Lord knows how long, the lock-up smells of the same things it always did: dry dust, old paper, WD40. As these fill his nostrils, they unlock so many memories: some that bring a wicked smile to his face, others a wince, and one or two that would have his cheeks burning with regret were he to dwell on them. Memories of past deeds, memories of a person he once was and hadn't thought he could be again. However, the lady had been very persuasive.

'Let me talk to you in a language I know you understand,' she had said.

She even came to see him. He liked that. No high-handed summons to her office. A person of rank and importance yet she knew you had to be humble when in need, especially given what lay in the past.

It wasn't purely about the money, either. She appealed to his sense of obligation too. He was a man of civic responsibility these days, weren't he? Who was he to refuse a respectable woman, one of some standing, in her attempts to catch a bank robber?

Plus there was the irresistible bonus attraction of finally putting a net around the one that got away.

'I need someone who understands what he's up against,' she said.

Albert was the last man she needed to tell that Innez had thus far proven rather tricky to apprehend. That was why he and Mr Spank would have to bring along a special friend who helped out on such risky occasions: Madam Boom. A right saucy piece, with a wicked mouth on her, though you wouldn't want a blow-job from those pouty little lips.

New handcuffs too. Forewarned is forearmed, after all. You live and learn.

*　　　*　　　*

Nine-millimetre Beretta M9 semi-automatic pistols: two. Fifteen-round magazines: ten. Heckler & Koch MP5 sub-machine guns with night-sight scopes and laser-aiming attachments: two. Thirty-round magazines: twelve. Aircrew Survival Egress Knife with calf-mounting sheaf: one. Ka-bar knives: three. Waist-mounting multi-sheaf for same: one. Hicks G12 grenade launcher: one. Teargas grenades: eight. Groeller-Duisberg gas-propelled dart pistol—

Beep.

The Guarantor instantly interrupts his inventory to consult his PDA. He glances at the screen very briefly then returns his attention to the flightcases in the boot of his Audi A8. The information is important, but it is not the message he must continue to wait for. The PDA shows him the name and location of a private airfield near Hereford, being where the handover will now take place. His contractor, Bernard, will be supervising the exchange personally, arriving via the Proprietor's private jet, which will be used to take the target out of the country. After that, the next, and final, leg of Darcourt's journey will be by sea.

The Guarantor completes his inventory and keys the airfield's detail into his satellite navigation device. He will drive there now, familiarise himself with the approaches, memorise the routes to and

470

from all major trunk roads in the vicinity, as he does not yet know from which direction he and his cargo will arrive. The last communiqué put him in a holding pattern, still awaiting the update that would inform him precisely where he must pick up his target.

There is a major police operation under way; he has been instructed to stand back and allow them to do their job, after which, the police will have far less control than they assume. Doors will be opened and actions taken by people who do not know for whom or why, or perhaps by people who suddenly found they had a compelling reason to cooperate. The Guarantor doesn't know the details, doesn't know who else is involved, any more than they would know about his involvement unless and until it becomes necessary. Neither he nor anyone will be given information that could prove in any way a burden or a risk.

Blind Complicity Measures, they are called. These are often a means of reducing the need for bloodshed, sometimes even a knowing trade-off by individuals in positions to make such compromises. The Guarantor's instructions today, however, contain no directive of restraint. He is authorised to eliminate all complicating factors if necessary, including those facilitating his access. His remit is to deliver Darcourt, an objective he implicitly understands to supersede all consideration of life and limb, not least his own. That is what makes him the Guarantor.

That is the deal when you are working for the Proprietor: you are paid extremely well, but on certain jobs it is mutually recognised that you will go to any length to deliver success. If you fail but survive, then you better play dead, because the Proprietor cannot afford it to be known that anybody has defaulted on this arrangement. That is why the apprehension of Simon Darcourt is being given the Proprietor's highest priority, and why the Guarantor will deliver him, or die trying.

CONSTABULARY ET ACETABULARII

It's party time, and Angelique has dressed for the occasion. She's gone for the classic LBD, a diaphanous but not clingy sleeveless affair with a split up the left leg almost to her waist. So far, none of her colleagues has made any remarks or let out a wolf-whistle. She's not sure whether this is down to a sense of propriety given the occasion, genuine fear that she'll break their noses, or simply that she doesn't look all that sexy. Several attempts to glimpse her tits through the fluttering gap at the neckline would suggest otherwise, though perhaps the fact that all they spied was nipple-tape created a sour-grapes effect.

She's not really sure the frock is her, to be honest, short notice and limited choice dictating her wardrobe. She had to find a dress that would meet certain requirements, and which they had two of in stock—not to mention the perennial issue of her size. She wasn't aware of ever feeling like she was surrounded by towering females, yet the racks of every fashion store she visited were stocked to cater to this army of Amazons. Still, when it comes to the little black dress, they say it's the accessories that really make it. Tonight Angelique is sporting a Walther P990, two spare clips of .40 calibre ammunition and a number of other little trinkets, all secreted about her in locations that dictated the need for the aforementioned high split and loose neckline. It's why she couldn't go with anything too figure-hugging, as it's not just her nipples that are taped down across her front.

473

To complete the look, and as an insurance against it getting chilly, she's also draped herself in a pashmina, one that can really bring the heat.

The doors are opening in a few minutes. Watches are synchronised, radios and earpieces tested. Shaw takes a long, worrisome look around the place, from the balcony to the stage, from the gantries to the exits. They've had complete control of the venue for two days, but this is the point when all the variables enter the equation: around three hundred of them, in fact, none of whom are aware of what is really about to take place.

Panic is the greatest threat. As well as a high-visibility presence of uniforms, there will be undercover cops everywhere, though when they get the signal to move in, their job will be more about controlling the audience than apprehending the villain. With any luck, if the Tivoli staff can get enough champagne down the guests, they'll be too pissed to realise it's not part of the show.

Shaw gets a signal from the stage, and another from front of house. The music will be starting in one minute, the doors opening in two. He turns and addresses the cops before they disperse amid the incoming audience.

'These people you're going to be among: they don't know it, but they are his allies tonight. They are the biggest danger to this operation's success, so be ready for chaos, be ready for desperation, be ready for abject stupidity. Just for fuck's sake don't shoot any of them, not even Steve Wright.

'I don't care how much of a sham this is. I want nobody to lose their concentration for one second. The fact that we're here to arrest a guy who's voluntarily giving himself up means we should be

474

all the more ready for surprises. Just because we think we know what the game is, doesn't mean we really know what the game is. Remember that.'

*   *   *

'Good evening everyone, and welcome to a very, *very* special occasion: a star-studded occasion, an exciting occasion, a glamorous occasion, but also a solemn occasion. Most of all, though, it's a defiant occasion. Because tonight, this is Jessica Hanson, once more talking to you live from inside the Tivoli nightclub in the West End. I was here presenting for PV1, the night Simon Darcourt hijacked the Nick Foster Lifetime Achievement party. A lot of the people sitting here right now were present that night too. We know we can't turn back the clock, but we can pay tribute to the stars we lost that night, and we can show our solidarity with the ones still missing. Most of all, though, we're here to tell Simon Darcourt this: the Luftwaffe couldn't do it, Al Qaeda couldn't do it, so you are kidding yourself, matey, if you think you can make *us* afraid to party!'

And with this, the lovely Jessica turns and steps to one side as the members of re-formed Nick Foster aberration SWALK take the stage to kick off the show with their twenty-year-old number-one hit: 'That Precious Look'. The horrifically dated Foster-signature synth-brass melody booms out as they spurt from the darkened mouth of this ridiculous snaking tunnel arrangement that dominates the rear of the set. It projects through the floor-to-ceiling illuminated fibreglass backdrop spanning the width of the stage, though as it is the passage through which all of tonight's stars will make their entrance, I can't decide whether the symbolism is more phallic or vaginal. Thus I don't know whether to describe the spectacle as four cocks emerging from a giant twat, or four cunts coming out of a prick.

Once they've been hoovered back up it again, Jessica informs us: 'This amazing stage-set behind me was designed by Maximilian. The tunnel—or tube might be appropriate—represents the distance between the real world and what's on the other side of our TV screens. But there are five people out there somewhere tonight who know that the world on the other side of that screen is just as real as the one the rest of us live in, so the tunnel also represents a passageway between the hostages and ourselves, a passageway to freedom we want to lead them through.'

She's *so* got this off a press release. Who the fuck is Maximilian? She says it as though we're all supposed to know the name, and yet I can tell she doesn't have a fucking clue either. Say a name with sufficient gravitas and people will nod sagely, going along with it for fear of betraying that they've never heard of the bastard and are thus ignorant philistines, or even worse, *not cool.*

Good for Jessica, though, you have to say. It's a nice touch, getting her to present the whole thing: a bit of a step-up from her role last time, being a glorified autograph-hunter for PopVision 1. Her star has risen a great deal since then, in fact, a serendipitous benefit of simply being the woman who was in the right place at the right time. Some of her footage that night is already legendary, inextricably bound to the moment this all began in the public memory, with her inadvertently ironic remarks, such as: *Bit of gossip for you folks, just between you and me, the rumour around Tivoli right now is that the guest of honour might be fashionably late for his own party . . . and when he finally makes his entrance, you can guarantee it'll be a show-stopper.* Yep, that little prelude is bound to make the next all-time-TV-moments pantheon, alongside 'I counted them all out and I counted them all back' and 'Mark, help me out here, say something'.

Her reward is that she's off PV1 and on to ITV. Off the pop beat too: her role as unwitting witness to tragedy has somehow earned her a bucketload of instant gravitas and she's landed a gig as a serious presenter on *Tonight with Trevor McDonald.* Not so serious as to say no to that *FHM* bra-and-panties photo-shoot, right enough, but her new credentials lend just that precise note of solemnity she alluded to for a night like this.

While I'm on the subject, it ought to be acknowledged that quite a few people have done very well out of me recently. I do hope they bear that in mind after I'm gone. I walked in right behind one, as it happens: Pippa Kimble, former washed-up ex-member of Sunshine Seven but now the girl who's got Jessica's old job on PV1.

Yes, that's right: I walked straight in, up the red carpet, in fact. Anything else would hardly be proper, now would it, what with this bash being all about yours truly. Nobody recognised me, which would have come as a crushing blow to just about everyone else who slinked along the same rouge rug this evening, but hardly a surprise to me. I did have my head all but welded to this TV camera at the time, after all.

As per my little arrangement with DCS Shaw, I was sent electronic versions of event tickets and various passes to laminate and tag about my person. This was so that I could sneak in as a guest, event staff or media personnel, with Shaw's side not knowing in advance which to look out for. Had to make certain things look authentic on the part of the cops, particularly the big moment of surprise, and method acting is always the most naturalistic. The multiple passes also created a miniature deniability firewall for the authorities, rendering it impossible for anyone later to definitively trace back my route inside and apportion blame.

I took all of the tickets, cards and laminates with me, but just as an experiment, I opted not to wear or present any of them: instead I dangled only a tiny press pass from the camera, knowing that it was the camera itself which would establish my credentials. Obviously I can't know for certain whether I was silently clocked by the police and allowed to pass, but I sincerely doubt it. As I stated, most of my head—and all of my face—was obscured by the camera I was lugging on my shoulder, and despite having no visible credentials, I just glided in behind Pippa, tracking my lens along the faces of the sad sacks lined up behind the cordon. Nobody asked to see any ID, and I'm not sure anyone even noticed the tiny card attached to my hardware. Pippa's own production crew never even cast a glance my way as I tailgated through in her slipstream.

I was allowed to roam unchallenged like a wandering holy man through a shrine or a temple. No celebs without telly, no telly without cameras. I was carrying the power to bestow the magic blessing, the sacred artefact that confers the mystic consecration. Let him through, let him through.

I notice Jessica has dropped an octave and altered her speaking-to-camera expression since she hitched a ride on the Trevornaut. It's only been a matter of weeks, but she's gone from sub-Kate-Thornton airhead bubbliness to full-blown Kirsty Young gravel-voiced austerity, complete with that burdened look like she just swallowed someone's come and is starting to have second thoughts.

As she witters on ungratefully about what a bad article I am, I turn to pan around the audience with my camera, and I'm almost moved to tears to see everyone who has turned out for me. There's the veterans of the Nick Foster night—Sandi Bay, Surfs Up!, Angel Cakes, Wendy Clear et al—and it's particularly touching that they came back, especially in the face of what traumatic memories this place must hold, being where they all watched their conduit to a possible comeback go literally up in smoke. However, it's some of the other faces that really mean the most. I see the Cadzows and Cassandra, and I bet none of them had to be asked twice, not least to put behind them the frightful matter of the recent misunderstandings. 'We honestly just took off for a few days . . .'

And look, look: there's Matt Willis, there's Steve Wright, and there's Jordan, with thingummy, her sun-tanned dildo. There's Vanessa Feltz, there's Linda Barker and, can it be? Yes! It really is Jade Goody!

Dotted elsewhere I spot Tamara Beckwith, Gary Rhodes, Gillian McKeith and, my God, Desperate Spice herself, Geri. Oh, and is that Amy

479

Winehouse next to her? Not sure: it could be Ozzy Osbourne—the table is quite close to the back, after all. I can also see Pete Doherty, Brian Harvey, *all* of Take That (though sadly no Robbie—oh for a *rapprochement*, boys, could you not find it in your hearts?) and many, many others.

Looking at all these famous faces is when I truly feel the pity of having so little time left. I'm like Oskar Schindler in reverse: I could have taken more, I could have taken so many more.

Before I can moisten up too much, my attention is drawn back to the stage. They're playing my song. Not *my* song, I should clarify: not 'Hurts Like Dynamite', but a less-than-golden oldie intended specifically for my ears. It's a dance-beat enhanced version of 'Funny How Love Is' by The Arguments. What do you mean you never heard of them? What planet have you been living on? They scaled the ear-popping heights of number eleven on the UK Indie chart back in the Eighties with their one and only release, a jangle-pop cover of a minor Queen album filler. It was a track that couldn't have been any more insipidly flimsy had it been released on flexi-disc, all the better so that it took up even less space in the dustbin of pop history.

The only reason it has been excavated from the great rock'n'roll landfill site is that The Arguments were formed by the remaining members of the first band I was in as a student, and they had this towering moment of triumphant success (did I mention number eleven on the Indie charts twenty years ago? When does the comeback stadium tour begin?) after I had departed. I'm interpreting that this is assumed to bother me somehow, as well as recognising it as my cue. It certainly tells me that de Xavia has had a strong hand in organising tonight's festivities, an observation confirmed by the accompanying sight of four figures dressed identically in Rank Bajin outfits, dancing in a barely rehearsed apology for choreography. They're all in black, wearing cloaks and cowboy hats just like the cartoon, their faces obscured with black drapes, on top of which are printed the character's iconic staring eyes and grid of teeth. It's so crass I almost want to applaud.

*   *   *

Angelique watches the hastily recruited 'Black Spirit Dancers' take the stage, but from then on she's back to scanning the audience, watching for a response. The houselights are higher than would normally be the case, but nobody has clocked Darcourt yet. Shaw had hoped they might huckle him quietly and with minimum fuss, aware that the other aspects of the deal would still play out, but Darcourt is clearly determined to have his moment literally in the spotlight.

She is standing by the wall, keeping a clear route between her position and the stage. She's half a dozen yards away from the nearest table, but close enough to see the expressions of puzzlement and distaste on the faces of those not sold on Jessica Hanson's preface that the routine represents a stiff British two fingers to the enemy. She knows she'll hear Shaw's voice in her earpiece any second too, and he doesn't disappoint.

'You've outdone yourself, Detective Inspector. Definitely the last time we let you play Harvey Goldsmith.'

481

Angelique had been given the job of arranging part of proceedings, specifically the section of the programme referred to among the police as 'the prelude', of which this was intended to be the climax. Her remit also allowed her to contract out the set design to one 'Maximilian', with the strict proviso that there be only one conduit between the stage and the secured area behind.

'We were told to put on a Get It Up You Darcourt party,' she reminds him.

'And could we have made it any more tasteless?'

She slips a hand inside the split in her dress and curls her fingers around the stock of the Walther.

'I've got a post-watershed moment lined up right here, chief.'

Angelique can feel the anxiety dissipate, the closer she gets to taking matters into her own hands. Her fear lay in her ignorance and helplessness, not to mention her conflict about undertaking this level of deception, but Zal has taken away all of those things. She's still jangling, but that's simply impatience, and as it all suddenly begins, it turns into sheer edge.

She clocks him just before he makes his move: guy with a big, shoulder-mounted television camera. There are several other cameramen in the place; Darcourt not only demanded it be televised, but sold it to them as a benefit on *their* side of the deal. This one suddenly stands out because he's walking across in front of the stage, his lens not pointed at the show or the audience. He then deposits the rig on the platform and leaps up to crouch alongside it, at which point the sides of the camera drop away to reveal a cluster of plastic explosives arranged cylindrically, wires and detonators daisy-chaining in and out of the arrangement.

Darcourt stands up straight, a few feet from the edge, holding out a blinking detonator at arm's length in his right hand. The dancers are the first to notice and respond, scattering in retreat, their Rank Bajin hats and cloaks hurriedly dumped as encumbrances to their flight.

The audience are still trying to decide whether this is part of the performance when Angelique draws the Walther and shoots him through the hand.

Cops just materialise from the walls in her peripheral vision, moving in to prevent a stampede, others heading for the stage, guns drawn. Darcourt has dropped the detonator and is holding his hand up, shivering and looking incredulously at the wound. Angelique changes her angle slightly, takes aim a second time and shoots him again, this time taking off his two middle fingers.

This is when the screams go up, none louder than his, while Raymond Ash's student band continue to provide their one-fingered salute over the PA.

Two cops—McGhee and Leitch—ascend the stage, while down among the audience their colleagues block the exits and urge everyone to remain in their seats. They get Darcourt face-down and cuff his hands behind his back, before marching/dragging him through the tunnel. Angelique picks up a cloak and three of the hats then walks briskly through the passage behind them. She pushes her way through the black drapes covering the backstage mouth of the tunnel and drops the garments on the floor as the music is abruptly cut off.

Backstage, Darcourt is deposited on a chair and his cuffs undone while a bandage is hurriedly located from a first-aid box and wrapped around his profusely bleeding mitt. Shaw stands against another chair, arms folded, a laptop at his feet and his eyes fixed intently on the monster they have landed, doubtless ruminating darkly that the real hunt has yet to begin.

'She shot me,' Darcourt splutters furiously, his face sweaty and mottled in his pain and shock. 'She shot my fucking fingers off. Why did you shoot me, you fucking bitch?'

There are three answers to this question, over and above the reason she actually gives him. One is that she wanted to know whether he would desperately try to recover the digits and thus indicate that a fingerprint ID sensor might be required to stop the gas and release the hostages. Another is that she needed to render him unmistakably identifiable. And the third is simply that she wanted to hear the bastard scream.

'Had to make it look authentic,' she replies. 'A hundred million costs a lot of pain, arsehole, and that was just small change.'

Darcourt looks at her with a venom more chilling than she has seen in two human eyes her entire career.

'A hundred million buys you five hostages,' he says. 'But it can just as easily buy four, and you'd still have to pay. Maybe I could even force you to choose who gets sacrificed. How would you like that responsibility?'

'Danny Jackson,' she replies. 'In a heartbeat.

You called that right, nobody cares about him. Next question.'

'It could also buy just one,' he reminds her. 'Now where's my fucking money.'

The bandaging complete, Darcourt is cuffed again, in front this time, as Shaw brings forward the laptop. It doesn't look like Leitch would make much of a nurse, but nobody's exactly moved by the wounded man's plight.

Darcourt is allowed to log on himself in order to verify that he is looking at his account and not some façade cooked up by the police's geek squad. As arranged, it is only now that he reveals the number, IBAN and Swift codes of the account he wants the money transferred to, preventing the authorities from taking advance measures to freeze it. Shaw makes the call to confirm that they have Darcourt in custody, and a few moments later, the transfer is approved. The British government, which does not deal with terrorists or hostage-takers, oh no, not us, matey, has just paid its most wanted criminal a nine-figure sum.

Shaw snaps shut the laptop and crouches to put it back in its carry-case. Angelique swallows, feels the adrenaline build. This is it. She checks her weapon, glances at the positions of the other cops in the room: McGhee by the door, Leitch standing behind Darcourt's chair. Both have their guns holstered. She runs a finger under the pashmina and takes a breath, readies herself to cross that final rubicon into the realm of deception. With one move, she will divert the course of events from the plan Shaw agreed with Darcourt, to the path picked out by Zal.

With Shaw still knelt on the floor, she draws her

Walther and levels it at the back of his head.

'Hands in the air, all of you, right now,' she orders.

'What the—'

'I *mean it*,' she says, pressing the barrel against Shaw's nape to drive the point home. 'Both of you: against that wall, hands up high. Chief, turn around, very slowly.'

'De Xavia,' Shaw demands, 'what the fuck's going on? Have you gone native from hanging about with that bank robber?'

'I'm sorry, sir. One day I hope you'll understand. No time to explain, but right now Mr Darcourt means more to me than the lives of anyone else in here.'

With this, she pulls away the pashmina and lets it drop to the floor. Shaw's jaw drops only slightly shorter at what is revealed beneath.

'Quite a necklace you've got there, hen. More bomb street than Bond Street.'

Angelique pulls out the detonator that has been strapped just below her left breast and holds it in her right hand.

'McGhee. Leitch. Put your guns on the floor slowly and slide them over there, away from the prisoner. That's it. Now take out your cuffs and secure yourselves to those pipes.'

As they comply, she removes Shaw's own cuffs from his inside pocket and attaches one to his right hand, one to her left.

'Radio your men and tell them all to holster their weapons. Warn the snipers outside too. I take my thumb off this button and the explosives blow. If they drop me, I drop this stick and we both die. DO YOU GET ME?'

486

Shaw nods. 'Yes, yes, fuck's sake. This is madness, Angelique.'

'Get on the fucking radio.'

As Shaw relays the order with his free hand, Angelique points the pistol at Darcourt.

'On your feet, bawbag. Go and get those hats and that cloak lying on the floor. Put on the cloak and a hat and bring me the other two.'

Darcourt reminds her nauseatingly of Gollum as he scuttles cravenly off the chair and willingly does as he is bidden while it serves his agenda.

'Now, put one on Shaw,' she commands. 'Backwards.'

Darcourt obeys, very gingerly protecting his wounded hand from unnecessary contact as he does so. The rear drape covers Shaw's face, effectively blindfolding him. Finally, she puts the last hat on herself, tucking the drape out of the way for the moment so that she can see better. She pushes Shaw forward towards the tunnel, beckoning Darcourt to follow with a wave of the Walther.

'What's going on?' Darcourt asks.

'It's your lucky day, Fingerbob. I'm getting you out of here.'

*　　　*　　　*

'. . . Hanson, speaking to you live for ITN from inside the Tivoli nightclub in London, where I am supposed to be presenting a tribute evening for the victims of Simon Darcourt. However, as you can see, the stage is empty and the event has been abandoned, but the dramatic news is that for Simon Darcourt himself, the show is over. A few minutes ago, a man subsequently identified by police sources as being Darcourt, leapt to the stage carrying what is believed to be an explosive device. At this point, the nightclub just came alive with police officers, so we can assume that some sort of undercover operation was in progress. One of them shot the intruder, and he was very quickly bundled backstage through that tunnel you can see on your screens.

'Now, the latest we've been hearing from the . . . no, wait, something must be happening. The police look like they've been alerted to something. They're drawing guns again. Some of them are gesturing at people to get away from their tables. It looks like they're clearing a passage, but surely they can't be bringing their prisoner back out through the club? There is someone coming from the tunnel, though . . .'

The Guarantor maximises the window on his laptop to get a better view. He sees three figures in matching hats emerge from the tunnel, single file. The first is a man in a grey suit, his face covered. He's moving slowly, one hand out in front: evidently, he can't see. Directly behind him—handcuffed to him, in fact—is a female in a black cocktail dress, her face covered by one of the veils he previously saw on those dancers. She is carrying a gun in her free hand and some other metal device is gripped in the cuffed one. There are explosives draped around her shoulders, which will be why the cops don't shoot her. Finally, falling in at her back so close as to be using her for cover, is Darcourt. He's got the same mask on, as well as a cloak, but even on these streamed TV pictures, the Guarantor can make out that one of his cuffed

488

hands is wrapped in a blood-soaked bandage.

The camera tracks them as they progress cautiously through the nightclub, cops lining their route, some training guns on the trio, others engaged in restraining the audience.

The Guarantor toggles his radio through the police bands, listening for their response. A minute or so later, he hears that the three of them have boarded a dark blue Ford Galaxy, driven by a fourth person, also masked, and are proceeding west, trailed by eight police vehicles and a helicopter.

He folds up the laptop, puts the A8 in gear and pulls away.

CHASE THIS LIGHT

Well, isn't Mademoiselle de Xavia quite the duplicitous little minx? Seriously, I'm starting to believe I might have been giving myself an unnecessarily hard time over Dubh Ardrain, as all my previous deconstructions of the fiasco failed to factor in just what a resourceful and audacious opponent I was up against. Not that I'm about to forgive her for shooting off two of my fucking fingers, but at least I now see there was a reason behind it, beyond gratuitous injury. I had to leave the things, knew in an instant that I'd lost them for good. The digits and blood vessels would only remain viable for a matter of hours. There's no time and no point thinking about getting them stitched back on.

Never mind, there are greater things afoot, such as the real revelation of de Xavia's worth, and it's not in pointing a gun at her boss, because any desperate nutter can manage that. No, I'm walking behind her and Shaw, about halfway down the tunnel when, just as I'm thinking she's dragging us all into some foolhardy and potentially catastrophic siege scenario, the whole picture changes in a twinkling.

A panel silently opens in the wall of the passage, and from it emerge two people, both wearing Rank Bajin headgear. The aperture appears behind the point Shaw has passed, though he isn't seeing much anyway with that cloth over his face. The first to step through the gap is a woman in a dress identical to de Xavia's, her doppelgänger outfit completed with gun, detonator and explosives (presumably as fake as the ones in my TV camera). She walks silently in step with de Xavia for the briefest second, during which some swift and subtle business takes place that leaves the newcomer handcuffed to the oblivious Shaw in de Xavia's stead. Meanwhile, behind them emerges a man in a cloak matching the one she had me wrap myself in, his hands cuffed in front, the right one wrapped in a red-drenched bandage. I now understand that she wasn't being entirely facetious when she said she shot me to make it look realistic.

The three of them proceed out towards the stage as de Xavia gestures to me to step into the aperture. She slides the panel back from the inside, and no sooner has it locked into place than I feel a lurch as we begin to descend on a platform, gliding smoothly but rapidly beneath the stage.

We both step off into a low-ceilinged room, one with thick walls going by the way the sounds of our footfalls are muted. It's dry and dusty, bare but for some boxes of tatty Christmas decorations. There's a door at one end, in the direction of the backstage area. De Xavia says nothing, just gestures with her gun to stay put for a moment.

I look at my watch.

'I would remind you that the gas is released in eighty-two minutes,' I tell her.

She shushes me. I wonder what she's listening for, then realise it's her earpiece. She's monitoring the response to the decoys above. We wait in place for a few minutes, saying nothing. I imagine we have quite a few questions for each other, but it's not only the need for silence that prevents both of us from asking any of them.

She exhales suddenly, like an athlete psyching herself for her next feat.

'Follow me,' she says, and who am I to argue?

She leads me through the door and into a narrow passageway. It snakes along through a couple of tight s-bends, suddenly emerging after the second of these into the dock where the props and scenery came in and out, a bare-brick relic of the venue's past as a Victorian music hall. The area accommodates a narrow iron staircase, a disused pulley system and a double-wide sliding door. It also houses a cop with a pistol, which he draws as myself and de Xavia emerge from the passage's last bend and into view.

'The fuck's going on?' he demands, flabbergasted.

'Put down your weapon,' she tells him, levelling her own.

'I can't let you do this, Angelique.'

'Put it *down*.'

'I *can't*. I have orders not to let anybody through this door, no matter who.'

'Put your gun down, officer.'

'You *know* I can't do that. Give it up and stand down before one of us does something we'll regret.'

De Xavia doesn't reply for a second or so. The two of them remain locked in the stand-off, maybe six yards apart, guns levelled. It's long enough for me to think my little reprieve might be over. Subterfuge and fake explosives are one thing, but we're talking real bullets now.

'I'm sorry,' she says. Then she shoots him three times in the chest. Blood sprays the sliding door as he is thrown back against it, before slumping, face-down, to the ground.

'Nice shootin', Tex,' I remark. She doesn't like this.

'Fuck you. There'd be a whole clip with your name on it if I didn't need you alive,' she assures me, her voice breaking a little.

'Well, *the police* want me alive, but I think you may have just resigned from the force. So are you going to let me in on your agenda?'

'You'll find out in good time,' she says, hauling open the door. There is a police squad car outside, sitting in a rain-washed yard off a side-street to the rear of the nightclub. 'But for now, the deal is the same. I want the hostages and you want five minutes' access to a laptop. Get in.'

She opens the driver's side door for me. I hold up my cuffed hands by way of indicating I'm neither free nor fit for driving.

'It's an automatic. You know where you're going. Take us there.'

'But my hand,' I remind her.

'Boo. Fucking. Hoo,' she growls, climbing into the passenger seat. 'Drive.'

It's fucking agony every time my right hand touches anything, so I grip the steering wheel with my left, pressing my right wrist against the other side to steady it. The indicator stick is to the right, but I'm fucked if I'm bothering with that.

De Xavia notes this when we stop to turn right on to Theobald's Street: a car gets blocked behind us as the traffic in the left-hand lane whizzes past. The driver honks his horn in protest at the lack of notice.

'Try not to draw attention to us, would you?' she says.

493

'I'm in a police car,' I remind her. 'They never fucking indicate.'

'No hand signals either,' she mutters acidly, the bitch.

She switches on the police radio, scans the frequencies to pick up the chatter. The decoy car and its train of pursuit vehicles are heading west along Victoria Embankment.

'Who were your accomplices?' I ask. 'Your boss said something about you hanging out with a bank robber. Obviously people who owe you big-time, given the fall they'll be taking.'

'They've got complete deniability,' she replies. 'They're performance artists, showbiz wannabes. They had no idea what they were really doing. They thought it was part of the show, or rather a hijack of the show. They know they'll get a slap on the wrist, but they also know it'll be worth it, because they'll be famous by the end of the week.'

'Just what the world needs,' I remark. 'More fame-grasping nonentities.'

'Well, you can hardly complain. You're the prick who's creating vacancies in the celeb market.'

Theobald's Road becomes Clerkenwell Road, which becomes Old Street, before I turn left on to Kingsland Road at Shoreditch. For the last hundred yards, de Xavia has been tapping away at her phone, composing a text. She holds off and looks up intently, however, when I hang a right under an archway and into the courtyard of four interlinked Victorian warehouses, now converted into a light industrial complex. How, she must be thinking, could he be running this whole scheme from a place like this, housing at least a dozen businesses, a hundred-odd workers and no doubt couriers zipping in and out from dawn till dusk?

I bring the police car to a halt behind the only other vehicle in the courtyard, a large street-cleaning truck, its brushes tucked in tight to its underside and water trickling from the hose clipped to the rear. De Xavia opens the passenger-side door but I stop her before she climbs out.

'Watch,' I say, and reach for a thigh pocket. It's awkward with the handcuffs, and even more so that I have to use my left hand rather than my right, but I get hold of the remote and point it at my invisibility machine.

'How far do you think we'd get in a stolen police car?' I ask her, as the cylindrical rear of the truck swings open, revealing it to be cavernously hollow, and a ramp automatically descends to the ground.

I had it made in Estonia so that it wasn't traceable. It's wider than normal cleansing trucks, but the design and liveries are sufficiently accurate that you'd need to see a real one right next to it to notice. It accommodates a Mercedes limo, an Escort van or indeed a police car, with just room enough to squeeze out round the sides. The brushes and suction pipes don't work, but there's a reservoir for the hose so that it drips water authentically, a timer releasing a volume every ten minutes so that it always looks like it's recently been in use.

'Appropriate, don't you think?' I ask her. 'For cleansing the place of detritus.'

'Except the shite usually goes in the back, not the driver's seat.'

De Xavia gets out and I drive the police car up inside. I reach into another pocket and turn my phone off silent before climbing out of the car. She keeps a careful eye—not to mention the gun—trained on me throughout as I shuffle my way along the side of the vehicle and down out of the truck.

'It's a stick-shift,' I inform her as I open the cab, indicating my handcuffs.

'Need to make sure you don't crash when you're changing gear, then,' she responds.

We climb in and I start the engine. I put it into first and pull away. Back on Kingsland Road, I manage the first gear-change with a minimum of swerve, taking both hands off the steering wheel as briefly as I can. I just hope we don't hit too many red lights.

De Xavia gets her phone out again, resuming what she was working on before our detour.

'Who you texting?' I ask.

'You'll find out,' she grumbles darkly. She presses the Send key and folds the phone shut.

'I'm sure I will,' I remark, turning my head so that she doesn't see my face.

There's a red light ahead. I have to stop. It means I'll have to struggle all the way up the gears again, but it will be worth it, because it lets me take my eyes off the road and look at her expression when the sound of an incoming text chimes from my left thigh pocket.

It takes a while for the penny to drop. Christ, and I thought the woman was a detective. I can see the connections being made, the possibility presenting itself, being ruled out, ruled back in. I see incredulity, horror and, finally, appalled, gaping-mouthed acceptance.

'*Inform us once you have acquired the target, then your contact will make himself known to you,*' I say, quoting my final text to her, sent only a few hours ago. 'So, this is your contact making himself known to you. Hello,' I add cheerfully.

'You,' she breathes.

'Take it as a compliment. I needed someone capable of getting me out of police custody, someone who would risk everything and stop at nothing to pull it off. I chose well, it seems, other than you mangling my fucking hand.'

'If I'd known, it would have been your balls.'

'Oh, dry your eyes, it's simpler this way. You said it yourself: the deal is the same. You get the hostages, including your folks, and I get the money. You just don't get me for a trophy. Sorry to short-change the constabulary, but I never fancied HMP as a provider of palliative care.'

We drive the rest of the way in silence, de Xavia sitting there numbed as she contemplates the humiliating enormity of what a mug she's been taken for. It's just a pity she won't live to relate this embarrassment to her erstwhile brother-in-arms, Larry the Little Drummer Boy. I'll have to live without avenging myself on that little cock-stain, however. I've put too much time and effort—not to mention every last penny—into this, and once it's done I won't be risking what I've built for anything so insignificant.

Three years I've been working on this: planning, surveying, constructing, purchasing, learning new skills, researching, reconnoitring, patiently putting all of the elements in place. That's right, three years. It was the idea that germinated and grew on those visits to see my son: *The idea that didn't go away; the sense of untapped power and possibilities that grew and grew, while only the price never changed*. What seemed inescapable was that identifying myself as the author of my deeds would be at the cost of my own life. Then one day I realised there was a way of paying that price with a dud cheque.

I'm made up in latex right now, distorting and disguising my features to a quite unrecognisable extent. It takes a bit of preparation, but it's mandatory. I've worn the same face on all my recent excursions around celebrity-land. More significantly, I also wore it when I photographed myself for the plastic-surgery headshots I secreted in Lydon Matlock's medical file—along with those forged hospital documents containing the sad, sad news about the big C.

The truck has a bit of a wobble as I change down in order to turn right, about a mile outside Marfleet docks in Essex. De Xavia comes out of her fug to look askance as I direct us towards what appears to be a rusty corrugated-iron fence. I push a button on the dashboard and a section of the fence slides inside behind another, opening a gap wide enough to let the truck pass through. From there, it's a quarter of a mile down the single-track approach road to where my operations base sits in dry dock, a rusty old hulk of a container vessel. It's in no shape to face the high seas, but it will play its part in my sailing off into the sunset nonetheless.

It won't be cancer, but I will die tonight, much the same as I died over Stavanger. I won't be able to sink all of the evidence in the depths of a fjord this time, but there will be too many half-incinerated body parts to make individual identification remotely possible, including whatever is left of Nick Foster, Four Play, Darren

McDade, Wilson Gartside and the sad homeless fucker I selected because his age and build were a close enough match to my own. Simon Darcourt will die in a massive explosion, along with Angelique de Xavia, her parents and all the over-celebrated oxygen-thieves in the cells adjacent to them. All that will survive will be the farewell video I upload to the web before hitting the detonator. But just before the fireball engulfs the ship, a solitary figure will invisibly slip away. A new man, a reborn man. (Albeit reborn minus two fucking fingers.) Most importantly, though, a very, *very* rich man, able to provide handsomely for his dependents, as well as financing an extremely comfortable retirement.

THE ONE THAT GOT AWAY

The drape in front of Zal's face has bunched a little on the left-hand side, so he only has a clear view out of his right eyehole, but as he merely has to follow the figure in front, it doesn't matter so much. He sees the front doors ahead, being hurriedly held open by cops, the red carpet of the covered walkway beckoning beyond. It's pissing with rain, water pouring off the sides of the awning on to the hard-core of rubberneckers and paparazzi, who are being swiftly urged back by more police.

There's two cop cars slewed half on, half off the pavement at the end of the carpet, but beyond them on the road is a dark blue Ford Galaxy with blacked-out windows. Only the windscreen is transparent, enough for Zal to see another ridiculous Rank Bajin grin behind the steering wheel. Zal overtakes the handcuffed couple and slides open the vehicle's door, stepping clear as Shaw and his captor climb inside. Then Zal hops up into the seat behind them and the vehicle pulls away before he has even slid the door closed again.

Zal tumbles back in his seat from the lurch of sudden forward movement, his headgear slipping halfway off in recoil at the impact. He shakes it clear then reaches forward with his cuffed hands and pulls Shaw's hat off too.

'Reckon he bought it?' Zal asks him.

'Well, if he's got any doubts at all, Angelique will fairly dispel them when she shoots Steve Payton.'

'Will he be okay?' asks Angelique's body-double, DC Ishtar Mitra.

'He's got two layers of Kevlar under those blood sacs. We tested it, he was fine. Won't stop the bastard putting in for sick leave, though.'

The Galaxy accelerates into the London night, a caravan of police cars drawn along in its wake. After a few minutes, it descends into an underpass, where it pulls in, allowing an identical dark blue Galaxy with matched plates to take over leading the convoy. The final two police cars in the trailing procession stop alongside as the others ascend out of the tunnel.

'At least that should keep the fuckin' media out of our hair for the rest of the night,' Shaw says, sliding open the door.

Officers emerge from both of the squad cars as the Galaxy disgorges its passengers on to the tarmac.

'Where is she?' Shaw asks the first cop.

'Left the building less than a minute ago, sir. Heading east.'

'How's the tracking signal?'

'Strong and clear.'

Mitra detaches herself from Shaw then undoes his remaining handcuff. Zal holds his own hands up expectantly, but Shaw takes the keys from her and sticks them in his pocket.

'You son of a bitch,' Zal says, more in anger than surprise.

'Sorry, son. Just doing my duty.'

'This isn't over,' Zal reminds him. 'You gotta take me with you. I need to know she's all right.'

Shaw stares at Zal, each man finding his impenetrability reflected back.

'We both know what each other is,' Shaw says.

'And we both understand what each of us has to do.'

Shaw turns to Mitra.

'Ishtar, you go with DS Thistlethwaite. Mr Innez and I will travel with DC Michaels.'

'Yes, sir.'

Shaw opens the rear door for Zal then climbs in alongside him. There's another cop at the wheel: Hewitt, Shaw calls him. Michaels holds up a small gizmo like a sat-nav device, a red blip flashing in its centre.

'Just turned into Theobald's Road, sir.'

'Let's move.'

The police car pulls away, the tug of its acceleration hinting at untapped power but its driver eschewing any conspicuous display of urgency. This would be a pursuit by stealth and patience, not haste. Zal's plan had gifted them that much.

* * *

'Have you thought of just asking them *real* politely?' he asks Angelique.

'Just working on the improbable hypothesis that your ingenious first suggestion has an infinitesimally tiny flaw, would you by any chance have a Plan B?' is her indulgent but impatient response.

'No, because asking real politely is gonna work just fine.'

She stares intently at him, conveying that the jokey prelude part is now over, but Zal returns a look of insistent seriousness.

'Zal, you're freaking me out.'

'No, not yet I ain't. But it's coming, so you better hold on to that seat.'

'Tell.'

'What I couldn't get past was that Darcourt didn't need an out. He's giving himself up in exchange for the money, and that threw off how I was looking at it. But then tonight it hit me after I had sold a dummy to those two bums working for Holland. I got them to pay a guy a grand to find out where I'd be. The reason they call it 'selling a dummy' is that you don't get people to merely *accept* a falsehood—you get them to invest in it. And the more they invest, the more it costs them to get that information, the more faith they will place in it. Now, you tell me, what did it take to get hold of that medical file?'

'More than a polite please,' she says, before it sinks in. 'Jesus.'

'Simon Darcourt does not have cancer,' Zal assures her. 'And he doesn't want to trade his last few deteriorating months of freedom for a hundred million pounds. It's a heist. He wants to get *away* with a hundred million pounds, which means he *does* need an out. And I'm looking at it, detective.'

'Fuck me.'

'Believe me, if we didn't have a lot of work to do . . .'

'That sleekit fucking bastard. *He's* got my parents.'

'Guaranteeing him not only someone who'll pull out all the stops to deliver him from the police when the time comes, but also providing him with a source right at the heart of the investigation.'

'Jesus. *Sources who do not even know they are*

sources. He asked for regular updates, and I gave them to him. That's where the tip-off about Bouviere came from too.'

'Leading you to the crucial piece of information he needed the police to discover for themselves. The murder and break-in wasn't to remove files— it was to *place* them. Bouviere may never even have worked on Darcourt: the important thing was he was crooked, which meant it was plausible he had, as well as making it inevitable the files got impounded. And the greater the lengths you have to go in order retrieve them . . .'

'The more valuable you believe the information to be,' Angelique completes.

'Which doesn't just go for the cancer documents. That picture in there was left by him too, so it's a cert he looks nothing like it. However, it's also a cert he'll be looking *exactly* like it when he shows up at the Tivoli.'

Angelique then gets out her mobile and makes some calls. The first is to her parents' neighbour, who can't remember the precise model, but is pretty sure that when they left for their big trip, they were picked up in a black Mercedes.

The second, once she has sourced the number, is to Ruth Baker, who swears on her children's lives that it wasn't her who informed the police her husband had information on Darcourt.

'My guess is he spoofed an email address from her company,' Angelique says. 'He wanted us to hear about his vigilante job, get us to join a few dots regarding his feelings for his son, his new-found but still fucked-up sense of morality. Make us believe he was doing this because he didn't have long left and wanted to pass on a legacy. And I

504

guess that means the Baker sketch might be more accurate than I thought.'

'Not that that would worry him,' Zal suggests.

'Why not?'

'Because nobody's going to be looking for him if they believe he's dead.'

*   *   *

Albert follows the convoy from a few cars back, a practised method of keeping his pursuit inconspicuous. It dips down into an underpass, then the traffic slows up, leaving him stationary at the top of the approach. He can see the dark blue people carrier emerge at the top on the other side, the cops in procession behind it. He's about to try something that would make a taxi driver blush so that he can skip the snarl-up and flank the pursuit from a parallel road, but then he catches a glimpse of a flashing light from down in the tunnel and reckons, nah, hold up a sec. If there's one thing he's learned with this guy, it's that you need always be ready to assume things ain't what they seem.

The traffic starts moving again, and he rolls at a gentle pace into the underpass. He sees an identical people carrier pulled in on the left, two squad cars flanking it, a cop waving to urge the traffic through, and on the pavement, in the midst of a little pow-wow, stands the long-sought Mr Innez.

Oh yes, called that one right. Another act of legerdemain from the slipperiest mark he ever chased, but this last dodge hadn't fooled Albert. Seen the show once too often, mate. You live and learn.

He proceeds out of the underpass and finds a spot to pull in on the other side of the next junction, throwing a U first so he can monitor the exit. Patience, he tells himself. It's taken five years but it won't be long now. Good things come, and all that. Once upon a time, this geezer had altered Albert's whole destiny, and tonight was when Albert was going to forever alter his.

*   *   *

Michaels' gizmo tracks the signal out to Essex, unmarked cars criss-crossing the route to provide visual updates on the progress of what is now known to be an oversize cleansing truck. There's upwards of a dozen police vehicles within a half-mile radius of the thing at all times, two choppers in a holding pattern out of visual range. And Zal's sitting in the middle of it, handcuffed inside a police car.

The journey ends outside Marfleet, at the concealed entrance to a disused dry dock. Maps are consulted and alternative entrances sought, in case Darcourt has any early alert systems that would be triggered by tampering with his camouflaged gate. Cars are slewed across both lanes, roadblocks established a quarter-mile back in either direction to prevent any traffic in or out. From the radio chatter, Zal has heard that there are also two police launches on their way along the Thames.

A section of fence is removed about a hundred yards from the concealed gate, allowing six of the police cars to pass through. They proceed at a little above walking pace, to prevent engine noise. The vehicles stop in a wide arc fifty yards from the edge of the dock, where an ancient container vessel dominates the landscape in its inglorious decay. There are two gangways: one wide enough for vehicles, and a second for traversing by foot.

Cops silently swarm from their cars, most of them gripping pistols, some toting HK carbines. Zal allows himself a wry smile, knowing bullets will not be Darcourt's undoing.

Shaw leans across and produces the keys to the cuffs. He undoes the left one, but then threads it through the purpose-installed steel retaining loop between the front seats and secures Zal's hand once again. Shaw and Michaels climb out, leaving Zal chained in the rear while Hewitt remains at the wheel. Shaw then casts a glance back inside the vehicle.

'Officer Hewitt,' he says. 'We're going to need all hands.'

'Shouldn't I stay with the prisoner?' Hewitt inquires.

'No, as long as he's cuffed to that loop, he won't be going anywhere.'

'Very good, sir.'

Shaw leans inside and whispers as Hewitt exits the car.

'As I said, Mr Innez, we both understand what each of us has to do.'

Zal nods. 'Thanks,' he says.

'Good luck.'

*   *   *

From the second she set foot on board, Angelique has felt a sense of dread about this boat so repellent that she could picture Charon refusing to charter it, even in a pinch. It's dank, gloomy, creaking and cold, but it's the smell that really chills her. The deeper he leads her into its passageways, the more choking it gets, an unmistakable stench of death and decay. It becomes so thick and enveloping that she fears she's going to puke. Then, when they reach the source, she does.

He takes her past an open door to a metal-walled room, the light left on inside to illuminate eight corpses in various states of putrefaction, four of them headless. She figures three of the others for Darren McDade, Nick Foster and Wilson Gartside. No idea about the fourth, but he looks fresher and in better nick than his cellmates, particularly the ex-members of Four Play.

This hulk is a labyrinth, criss-crossed with walkways and corridors. She knows, then, that he took her this way deliberately, to goad and disgust her. She can't afford to get squeamish, though. He turns around and smirks as she wipes her mouth. Same old Simon: has to sing when he's winning. Or when he thinks he is.

She hears a crackle in her ear: Shaw, now in transmission range.

'Alpha X-Ray, we're in. Play ball. Make the deal.'

She had made a show of taking off her earpiece in the police car as they drove away from the Tivoli, but had a second one on the other side, disguised to blend in with her lobe. She's got a tracking device strapped to her chest, a wire taped to her midriff, and one other surprise lashed to her thigh.

They pass through into a spacious hold where she sees furniture she recognises from McDade's 'hotel room', the giant amplifier Foster was electrocuted on, McDade's gallows and the remains of the synths from Four Play's final video.

'Long as we're taking the tour, I want to see the prisoners,' she tells him. 'My parents first. And they better be alive, Stubs, because you don't need testicles to work a laptop.'

'I would warn you against doing anything rash, de Xavia, because I promise, I have taken every precaution to ensure that what harms me, harms the hostages.'

'Nevertheless, I want proof they're alive or I've no reason not to shoot you where you stand.'

'Patience, girlfriend. It's all part of the package, as agreed.'

They pass out of the hold into another passageway, stopping briefly at a dense iron door with a wheel-operated lock, which Darcourt turns and swings open. Inside she can see the common area with the big TV on one wall, and the sealed doors of the four holding cells around the room.

'*Celebrity Rescue*,' he states tiredly.

Further down the passage, he opens a second wheel-lock, giving on to a small hallway housing several more doors.

'Your parents and the saintly Anika are in these ones,' he says.

'I want to see them.'

'Then keep following the tour.'

Darcourt leads her down a narrow flight of stairs to a lower deck. Then they enter the heart of Darcourt's squalid operation. It is a room full of PCs, monitors, TVs, speakers, mikes, routers, cables and assorted baffling-looking gadgetry that would probably have Meilis spaffing his load. Darcourt hits some switches on a console and several screens flash into life, showing the interiors of each cell. She scans them all, keeping half an eye on Darcourt, but he doesn't seem to be planning any surprise attacks. Doesn't fancy the odds; not right this second, anyway.

'How do I know this is live?' she asks.

'You can talk to them.'

Darcourt hits some more switches and indicates a microphone.

She swallows, has to clear her throat just to manage these few words.

'Mum? Dad? It's Angelique.'

On the screen, they both rise from their trance-like despond, looking confusedly at each other like they need verification of what they are hearing, barely daring to believe it.

She's about to tell them to hang in, she'll have them out soon, when Darcourt flicks another switch to kill the feed.

'The clock is ticking,' he reminds her. 'Twelve minutes until . . .' he makes a hissing sound to indicate the gas. 'Plenty of time later for catching up, if our business is successfully concluded.'

'So how do we work this exchange? You were conspicuously hazy about the final details.'

'It will be my enormous pleasure to demonstrate,' he replies, and she believes him. He pushes a button on the console, causing a cylindrical panel on the wall opposite to rotate inwards, opening up a tube-shaped compartment. It's big enough for one person to fit inside, just, and its interior is blank but for three devices

510

vertically integrated into one side. These are, in ascending order, a keypad, an LCD screen and a second glass panel, hooded like a microscope.

'This is your guarantee,' she observes.

'Well, your own conduct tonight has proven I couldn't possibly trust the police. You even double-cross each other.'

'Time's wasting, Stubby. I want my hostages. What is this, a teleport?'

'A portal, certainly. The portal to your hostages' freedom, and to mine, mutually guaranteed. The keypad controls the locks on the cells, but it needs the scanner to authorise it, and the scanner requires the door to be locked before it will power up. In short, I can't release the hostages until I'm locked in there, clear of you and your gun. And I won't lie to you, once that door is locked, I won't be coming back out of it. There's an escape hatch in the floor, and I will be leaving through that.'

'Albeit handcuffed.'

'Thanks for the concern, darling, but I've made provision. I've been planning this for . . . a long time. I'm pretty sure I haven't overlooked anything.'

'A *retinal* scanner, did you say?'

'That's right. No codes and no passwords. Nothing that can be hacked, faked or coerced.'

'And what's my guarantee that once you're inside that chamber, you'll actually release the hostages?'

'It's in my interest to keep you busy. The timers and the gas canisters are buried in cement. I did that as soon as I set them. Even if you find them, they can't be stopped. This keypad unlocks the doors, but only for two minutes, and the cells need

511

to be opened from the outside. This was my insurance. I mean, don't flatter yourself, dear. You did well, but I couldn't stake my whole plan on your success. Had to make sure the arrangement worked even if I was surrounded by cops, though I wouldn't have told them about the escape part: they'd have thought they were trading their two minutes with the locks open for my two minutes with a laptop.'

She hears a burst of static, a prelude to Shaw's voice in her ear. He will be getting all this via her wire, so presumably he's about to confirm they're in position to make Darcourt's getaway a little trickier than he's anticipating.

Except that, on this case, you can't presume anything good.

'Alpha X-Ray, listen carefully,' Shaw says, his steady measured tone itself betraying the urgency and panic his voice is trying to conceal. 'We've been below decks and discovered this whole fucking rig is wired to blow. Repeat: this boat is one massive bomb. There's enough explosive down here to put this hulk into orbit. That's how he's planning to cover his getaway. You can't let him get into that chamber or he'll kill us all.'

Angelique feels her eyes popping, hopes Darcourt assumes his own revelations have been remarkable enough to provoke it.

'Any more questions,' he asks. 'Or do you actually want these hostages? Six minutes, by the way.'

'Just one,' she replies. 'Why are you telling me?'

He sneers, clearly looking at her with all the anger and disgust he's been storing up since Dubh Ardrain. She knows that the short answer is because she's about to die anyway, and he'd like to see her off with one last gloat. However, she indulges him to spunk out the long version while she readies herself for what she'll have to do.

'Because, de Xavia, even now, having betrayed your colleagues and murdered a fellow officer purely for *your* purposes, you still think you're a fucking hero, and I want to disabuse you of your delusion that you're on some kind of superior moral footing.'

'Superior footing? Darcourt, you're so far down the moral ladder that if you looked up with a telescope, you still wouldn't be able to see the soles of a paedophile's feet.'

'Well, let's just see where *you* stand on that ladder, shall we? Five minutes to go, time for a new deal. Truth is, I lied about the cancer. I'm not dying, nor planning to any time soon. So here's a moral conundrum for you, Angelique: you can kill me right now, at the cost of letting your parents and the other hostages die. Or you can save them, knowing it comes at the cost of me living to kill another day. Clock's ticking. It's your choice.'

Angelique raises the Walther.

Darcourt stares back at her, looking like this is just a bore.

She lowers the gun.

'No choice at all, was it?' he says. 'But don't beat yourself up too much regards the living-to-kill-again thing. I may just retire, what with that generous government pension to fall back on.'

'History won't be kind, Simon,' she warns, slipping her free hand through the split in her dress.

Darcourt begins walking towards his portal.

513

'*There are names written in her immortal scroll at which Fame blushes,*' he tells her airily. 'William Hazlitt wrote that.'

Angelique watches him turn to enter the chamber, then shoots him in the base of the spine. He flails against the wall and collapses to the floor, face down, bleeding, paralysed.

She walks across to where he lies, slipping a knife from her thigh. She crouches down, takes hold of his hair and lifts up his head.

'*Out, vile jelly,*' she retorts. 'William Shakespeare wrote that.'

THE INESCAPABLE

Zal considers it might be abusing Shaw's goodwill to steal his ride, but Hewitt did leave the keys in the ignition, and with there being cop cars, roadblocks and even helicopters everywhere, he figures it's the least conspicuous way to get out of there. He drives back out through the gap in the fence and on to the main road, then turns left into the adjacent dockyard a hundred and fifty yards before the western roadblock. He's guessing it won't look suspicious: just another car covering a part of the perimeter they may have overlooked. The gate is padlocked, which is perhaps why the police didn't venture in there already just for recon. Takes him about ten seconds, then he's able to drive on through.

The dockyard is about as derelict as Darcourt's place next door. There's a crane by the waterside, the metalwork so rust-eaten that it looks like one attempt to swivel it on its turntable would cause it to disintegrate in a big flaky brown cloud. For company it has three freight containers in only marginally better shape.

He ditches the car where it won't be visible from either the road or the dry dock next door, and scrambles his way to the summit of a shored-up embankment topped by the mesh fence separating the two properties. He crouches down and gets comfortable, ready to wait it out. The rain isn't letting up any and he'll be soaked pretty soon, but he ain't leaving until he's seen her come out of there safe and sound.

Zal has an elevated perspective upon the container ship and the dry dock, but the darkness and the rain mean there isn't much to see right now. Then he makes out some movement: a single cop down on the apron of weed-strewn broken concrete, who leans into one of the cars and switches on its headlights. In a matter of seconds, the cop has repeated the operation on all the vehicles arced around the ship, bathing the gangways with light. No issue of stealth any more, so the show must be over. Confirming this, he hears the sound of sirens and sees a quartet of ambulances on the main road, the cars comprising the roadblock reversing out of the way to clear their path. Meanwhile, some more cops are busy busting open Darcourt's sliding gate to allow direct access for the emergency vehicles.

Down on the bigger of the gangways, he sees the first figures emerge just as the ambulances reach the arc of police cars. They come out two by two at first, each pair comprising a staggering, enfeebled figure accompanied by a cop. Paramedics jog towards them, wrapping them in blankets and leading them towards the ambulances. He watches one guy emerge alone, shaking off an offer of assistance from a cop but then being urged to let the medics take a look at him. Zal figures him for Dale.

Still they keep coming, until he's also accounted for the TV presenter, the soccer player's wife and the asshole comedian. Then, proceeding more slowly and anxiously than the others, comes the talent-show kid, Anika, who all but collapses when the paramedics reach her. Still no de Xavia, though, nor Shaw; nor Darcourt.

516

Then two cops lead a couple of the paramedics forward with a gurney, which they roll up the gangway. Zal feels a brief pang of fear and concern, but only long enough to remind himself who he's waiting for here. There was only ever going to be one person sure to need carried out of there tonight: the one who fucked with Angelique.

And yep, there he goes a few minutes later. Hard to tell from this distance, but as they're holding some kind of pack to one side of his face, it looks like Zal was on the money when he predicted Darcourt would be using a retinal scanner. Yeah, you had your playtime, dude, but I guess your mom never warned you: it's all fun and games until . . .

Darcourt is wheeled into one of the ambulances, which takes off at speed, blue lights and sirens. It's followed by one of the squad cars, and once it hits the main drag, it picks up two motorcycle outriders who take the vanguard. Adios, motherfucker, you got off easy.

Zal turns his gaze back to the gangway and now, finally, he sees her. Her mom is in front, being helped along by Shaw, while Angelique herself has her arms around and her head tucked against her daddy. When they reach the bottom of the ramp, Mrs de Xavia ditches Shaw and turns. The three of them coalesce into one tight and tearful hug, from which they don't look likely to emerge any time soon.

She's whole again. They're all whole again.

Zal allows himself a sad smile, thinks of what might have been, but he can't dwell on that. He feels happy for her, and proud of the part he was able to play when she needed him, but he still has work to do, one last gift to give her. The gift of

517

leaving.

He takes a final look but can't see her face. She's still all wrapped up in her parents' arms, and that kinda says it all. He turns to make his way back down the embankment, but loses his footing on the scree and scrambles the last few yards on his ass. He's about to pick himself up when a figure emerges from behind one of the freight containers, brandishing a shotgun.

''Allo again, my son. Long time no see. Real blast from the past, this, innit?'

'Oh, shit.'

'And there was me worryin' you might not recognise me after all these years.'

'The cockney Bobba Fett. How could I forget?'

'Ah, now, see, I'm not really in that line any more, but I was convinced to pick up Mr Spank one last time by a lady who reckons she has some unfinished business regarding a bank job up in Glasgow.'

'Fuck. Don't you guys have a statute of limitations in this country?'

'I'm not a lawyer, mate, but I'm bloody sure there's no loophole's gonna get you off the hook on this one. She's got very strong feelings regarding this matter. Very motivated, if you catch my drift.'

'Is this you saying I could alter your motivations if I topped her fee?'

''Fraid not, my son. I'm a man of integrity these days. And you should never break a promise to a lady. Come on. This way.'

Fleet leads him to a high-sided van, opening the rear doors to reveal a steel loop welded to the ceiling inside. He then produces a set of handcuffs.

Zal can guess what's coming next; unfortunately he doesn't guess quite all of it.

'Clothes off, matey,' Fleet orders. 'Down to your drawers. See, I've been reading up on your sort. Got your clever little tools squirreled away somewhere, ain't you? Well, not tonight, my son.'

Zal peels off the wet clothes. He pleads to be allowed to keep his trousers as a token gesture of defiance, given that his hands will be chained up overhead, but he knows Fleet isn't for budging. Fleet keeps the gun pointed at him the whole time, carefully observing as Zal puts the cuffs through the loop and fastens them around his wrists.

Once he is secured, Fleet has a root through Zal's discarded trousers. He holds up a small wallet full of picks and taps the side of his head, flashing Zal a smug smile.

Oh yeah, you the man.

'You know, I really should be wearing a seatbelt,' is Zal's parting shot as Fleet closes the rear door. 'Wouldn't want you to get a ticket.'

To his credit, Fleet does drive pretty smoothly, and indeed slowly, which ensures that the journey seems to take an interminably long time. Certainly it feels that way when you have to keep your arms above your head the whole time, as Zal does purely for appearances' sake. He freed himself from the cuffs about half a mile in. The wallet was a decoy for precisely this kind of situation: he's got another set sewn into his jacket, but his real emergency picks he keeps in his wristwatch. Well, *duh*. Jesus, what good would something hidden in his trousers be when he's strung up like this?

Eventually the van comes to a stop and the engine quits. Sounds like they're inside: a warehouse or a real big garage. He hears Fleet close his door, then the sound of footsteps walking away. Okay. Zal is deciding whether to try and spring the rear doors or bust his way into the cab and hotwire the engine, when he hears more footsteps, this time coming towards the van.

Shit. He puts his hands back in position. He'll need to choose his moment very carefully, wait until Fleet is real close, and when he hits him, it'll have to be a knockout blow. Can't afford to take any chances with that shotgun around.

Zal tenses as he hears the rear handle being pulled. It jams a bit, takes a second creaking tug. The doors swing open and he finds himself looking not at Fleet, but at the woman who hired him to make sure Zal didn't get away.

She's smiling apologetically, sniffing away tears.

'You need a hand out of there?' she asks.

Zal nods. She climbs up inside, holding Fleet's keys. He waits until she's a foot away, then drops his hands and hugs her. She doesn't sniff back any more tears now, just lets them flow.

They hold each other for a long time, neither able to say anything. Then she reaches into her pocket and produces her phone, which she hands to him.

'I got a text from an unrecognised number, came during that press conference after Holland's men tried to lift you. That's when I knew what it was I needed to do.'

Zal looks at the screen. It says simply:

He loves you.
Don't let him leave.
Morrit.

LIBERATION

The Guarantor watches the ambulance approach through a detached night-vision rifle scope, zoomed to seven-x magnification. It's four hundred metres away, according to the range-finder, and in accordance with what he picked up on the emergency channels, it is escorted by two outriders and a car in the rear, heading for Millfield Hospital.

He has chosen his spot and estimates forty seconds to intercept. It's a two-lane stretch, a straight channel hemmed in on both sides by concrete walls. In the passenger seat sit the Hicks G12, the HK MP5 and the Groeller-Duisberg pistol, both Berettas holstered either side of his chest.

Thirty seconds. He pulls away into traffic, heading towards the convoy. Twenty seconds. He accelerates to a steady thirty and holds his speed. Ten seconds. The outriders are less than a hundred yards away.

Zero.

He brakes smoothly but rapidly and slews the A8 across the oncoming lane. Both motorcyclists slam into the side of the car, a fraction of a second apart. They are sent headlong over the top of the vehicle and skidding on to the tarmac. The ambulance fishtails its way to an emergency stop, the squeal of the police car's brakes audible behind it.

The Guarantor climbs quickly from the A8, the MP5 slung around his shoulders, the grenade launcher in his grip. He fires four teargas grenades over the top of the ambulance, engulfing the cops as they attempt to emerge from their car. Then he takes the MP5 and opens fire, shredding the police vehicle's tyres and ripping into the engine as the cops flee, blinded and choking.

The Guarantor slings the MP5 again and takes the GD from his belt, holding it in his right hand as he opens the ambulance doors. Inside he sees two paramedics next to a trolley, upon which lies the target. They both put their hands in the air, disorientated and terrified. The Guarantor shoots each of them with a tranquilliser dart straight to the chest. They look at him in panic and disbelief.

'Lie down so you don't hurt yourselves when you fall,' he advises them. 'You will survive. Assure the British authorities that Darcourt will not.'

The Guarantor wheels the trolley out of the rear and pauses only to put a dart in the driver before bundling his package into the back of the A8. He describes a tight arc, assisted by his handbrake, as he executes a u-turn, then accelerates away as fast as German engineering can facilitate.

He reaches the airfield in thirty-three minutes.

The plane lands thirteen minutes after that.

'He's been shot in the back and has lost his right eye,' the Guarantor explains to Bernard. Always wise to get the faults mutually acknowledged before one signs over the merchandise. 'Looks like they sedated him. He may have lost a lot of blood, so I don't know how long he'll live.'

Bernard hands the Guarantor his payment as two men carry Darcourt aboard the plane.

'Longer than he'll wish to,' Bernard says.

SPIRIT OF ATHENE, ONE MONTH LATER

The sorcerer sits in his chaotic and ramshackle study, slumped in anguish, his head upon his desk and his right hand clutching the edges of a framed portrait of his departed love. His grip upon it slips and it falls forward, but the portrait disappears in a plume of flame before it can hit the ground.

He gets to his feet and reaches for a dusty spell book upon his shelves. A moth flies from between its pages as he opens the tome, and with a gesture of disgust he waves three fingers at it, turning it into a dove. It flies around the room and lands upon a perch while he thumps the book down on the desk before placing a black crucible alongside it. Flicking through the pages with his left hand, he produces a shrivelled wand in his right and embarks upon a frenzy of spell-casting, causing various of the arcane and occult artefacts about the study to levitate, vanish, reappear or transform, his frustration growing with each clearly unintended feat. Finally, he causes a porcelain bust of his beloved to suddenly become liquid, the sculpted head disappearing in a milky splash, whereupon he collapses to his knees on the floor in ultimate despair.

At this, the study falls dark, allowing him to become aware of a glow outside his window. He looks up in astonishment and sees her face in the glass. As he climbs waveringly to his feet, the walls around him disappear, leaving him alone in the presence of his love, his departed, his angel. She is floating in the air against a backdrop of stars, her hair like an aura around her, her dress shimmering about her body like wings. The sorcerer staggers towards her, reaching out a hand. She beckons him, seeking his embrace. Then, as his arms are about to clutch her, she vanishes as though those wings, that aura, simply collapsed upon themselves.

He takes another few despairing steps through where she had been, then finds himself standing in front of what a spotlight suddenly reveals to be her headstone. He nods, understanding, accepting she is gone, and reaches into his coat for a flower. All he can find is a withered and sad-looking specimen. He places it upon the grave and walks away. As he does so, it becomes a bounteous bouquet of red roses, then the curtain falls.

A few moments later, Zal is bowing before tumultuous applause, but none of it sounds as good as when his angel joins him to take her bow. She's breathing hard as she stands next to him, smiling and exhilarated. He recognises that look: she's feeling the buzz, and he intends to get her dependently addicted.

Morrit is trying to disguise the fact that he's got a tear in his eye as they come backstage. This final illusion, the Vanishing Angel, was Zal's take on one known as the Mascot Moth, which the old man hasn't seen performed in decades, and which he had always wished to stage again. The old man is delighted to witness that even in the twenty-first century, it is still capable of astounding an audience. The angel disappears live on stage, right before their eyes: no billowing cloth to hide the magic moment, no cabinet, no smoke pellets. It was pioneered by deKolta and perfected by David

Devant almost a century ago, but it took Morrit's knowledge and instruction to bring it back to life. Not to mention the talents of a very promising new magician's assistant, who has recently opted for a radical career change.

Angelique was sick that morning, so nervous was she about making her stage debut, but as Zal told her, everybody went through that. He was sure she'd be great, and so far she's proven him right.

'It's much the same as your old job,' he'd explained. 'You have to perform all kinds of athletic and daring feats, except nobody will try to shoot you. You may have to dodge a few swords, and you will almost certainly be sawn in half, but there's a waiver covers that.'

Truth be told, Zal was probably even more nervous than Angelique about tonight, but not due to any fears regarding her performance. His principle concern was that he wanted her to love it, because he wanted this to work. From what he witnessed onstage, neither of them had ever had anything to worry about.

'What did I tell you?' Zal says to Morrit. 'She's a natural.'

'You sure are, pet,' Morrit assures her, pouring both of them a flute of champagne. Angelique declines, picking up a glass of water instead. Come to think of it, Zal hasn't seen her touch a drop for a while, figured she was avoiding anything that might affect her concentration. She's earned herself a drink now, though, surely.

Zal chimes his glass against hers as Morrit impatiently ambles off to fuss over the Vanishing Angel apparatus.

'You did it, kid,' he tells her. 'You were great. I

reckon you're ready for the big time.'

She smiles, puts out a hand and rests it on his thigh.

Zal takes a big gulp, feeling the nerves again in anticipation of what he has to say.

'Which is just as well,' he goes on, 'because we've been offered a very big opportunity. Singapore. Extremely good money, and it's owned by the same people as own several hotels in Vegas. Play it right, and one day I'll be able to take my show to the city where I first watched my dad perform.'

She gives him a weird look, one he can't read at all. It's like she's real happy, but she's also afraid she's about to let him down.

'Look, there's no pressure,' he assures her. 'This was on the strength of my one-man show, and that's what they're expecting. We can work the assistant stuff into the act gradually.'

She smiles and gives a subtle shake of the head, mumbles a reply.

'. . . a little late,' is what he catches.

'To start this? Don't be insane. You're only, what, thirty-six? And you're a natural.'

She laughs, takes hold of both his hands, pulling him towards her. 'I believe you,' she says. 'But it might be tricky, in the short term at least.'

'Why?'

Then she draws his head close, puts her mouth to his ear, and whispers.

'I didn't say *it* was a little late . . .'